Managing health and hygiene in the workplace effectively

Managing health and hygiene in the workplace effectively

James Stowe

London: The Stationery Office

YOU NEED THIS BOOK FIRST

ABOUT THE AUTHOR

James Stowe has many years' experience in the field of health and safety, as a Health and Safety Manager in industry, and as a polytechnic and university lecturer in Safety and Risk Management. He was a founder member and vice-chairman of the National Examination Board in Occupational Safety and Health, and was a long-standing member of the Council of the Institution of Occupational Safety and Health until retirement. He has written many articles on health and safety topics and is currently involved in consultancy work.

© The Stationery Office 2001

Applications for reproduction should be made in writing to The Stationery Office Limited, St Crispins, Duke Street, Norwich NR3 1PD.

The information contained in this publication is believed to be correct at the time of manufacture. Whilst care has been taken to ensure that the information is accurate, the publisher can accept no responsibility for any errors or omissions or for changes to the details given.

James Stowe has asserted his moral rights under the Copyright, Designs and Patents Act 1998, to be identified as the author of his work.

A CIP catologue recorded for this book is available from the British Library.

A Library of Congress CIP catalogue record has been applied for.

First published 2001

ISBN 0 11 7028266

Printed in the UK for The Stationery Office
73349 C20 11/01 9385 16027

Contents

The protection of employees from the health hazards of their occupation is of relatively recent concern. Although the principal statute dealing with this is the Health and Safety at Work etc., Act 1974, this is still widely thought of as concerning only *safety* at work. Occupational ill health was, and still is, often considered as an inevitable consequence of specific occupations, therefore it 'ran in the family'.

The contraction of the United Kingdom's heavy engineering manufacturing base has reduced the incidence of diseases associated with, for example, mining, smelting and refining. But many of the old sources of such diseases are still with us and new sources have been created by advancing technology. Respiratory diseases still feature prominently in occupational health surveys.

There is still widespread uncertainty as to what occupational health and hygiene actually entails, especially as the practitioner in this field is known as an occupational *hygienist*. A person qualified as such recognises that there will nearly always be toxic and potentially hazardous materials to be processed or handled at work and that some degree of physical stress such as noise, heat, radiation and so on will be encountered. Also, because many occupational diseases and illnesses are chronic as opposed to acute their symptoms develop slowly and manifest their worst effects when the victim gets older or has retired from work.

The Health and Safety Executive's contemporary slogan 'Good health is good business' is perfectly true. However, there are clearly defined limits to which an employer needs to go on the question of occupational health. Many seemingly well orchestrated representations on the dangers of asbestos, for example, seem to ignore the fact that if such asbestos in any structure, installation or building is completely sealed or encapsulated, it is perfectly safe and does not need ripping out, at great expense, causing production stoppages and probably spreading asbestos dust in the process.

This book strips the mystique away from occupational hygiene and tells employers in small to medium enterprises (SMEs) how far, and only how far they need to go in complying with the law. The concepts of 'Occupational Exposure Standards', 'Time-Weighted Averages', 'Maximum Exposure Limits' and so on are explained in simple terms. Examples of assessments of exposure to harmful substances under the Control of Substances Hazardous to Health Act are given and the employer is shown how to make sense of the masses of often unwanted 'safety' information provided by suppliers to cover themselves in law.

There are literally thousands of substances on the market of varying degrees of toxicity to which an employee *could* be exposed. Which of these are most likely to feature in an SME's work environment, and in what amounts? A solvent-based bottle of correction fluid poses an insignificant health risk compared with a large drum of the same solvent in a degreasing shop. This book clearly demonstrates how to implement health safeguards for employees but in proportion to the health risks posed by their work environment.

History tells us that the protection of workers' health was a concern of employers long before their safety became a serious consideration. However, there were some serious lapses in this concern in the UK during the early years of the Industrial Revolution. At the start of the 19th century the 1802 Health and Morals of Apprentices Act required, *inter alia,* factories to be 'cleansed by two washings with quicklime per annum and the admission of fresh air by an adequate number of windows'. This start on improving sanitary conditions in factories led to the movement to improve health in towns, the subsequent appointment of sanitary inspectors – better known to the public they served as 'rat catchers' – who today are local authority environmental health officers.

Between 1910 and 1912 modern occupational medicine was born. Physicians wrote about occupational diseases and the public was becoming aware of occupational health hazards. Long-needed legislation was finally beginning to emerge and the science of pathology was being applied to occupational health, particularly in the area of lead poisoning. As Great Britain led the world in industrial development it was only natural that British factory and health laws formed the first worker health related legislation in the world. The early twentieth century is, then, for all practical purposes, the beginning of modern industrial health legislation.

THE ROBENS REPORT

Prior to this report in 1972, legal requirements concerning the health of people at work were very selective in that only workers in manufacturing and ancillary industries, mining and quarrying were covered. Where they did apply they had a strong hygiene theme, i.e. they specified washing facilities, ventilation, provision of protective clothing or equipment and the keeping of premises clean by regular sweeping and removal of refuse. Quite often, only token observance of these legal requirements took place.

A major reason for setting up the Robens Commission was because in the post-World War Two era the development of technology, and society, had overtaken pre-war legislative controls. The 1960s saw an explosion of more sophisticated manufactured products made by production systems using higher risk methods and new raw materials and substances. Often little was known about the health risks of these factors; knowledge was restricted to the fact that they did a good job as far as production was concerned (it was to be two more decades before any thought was given to the effect these developments were having on the 'health' of the environment). As occupational ill health increased and new and unfamiliar diseases started to develop, apprehension about the risks the new technology was creating grew. Alongside the increase in work related ill health cases the number of accidents and injuries at work were mounting and the extra load this was putting on the National Health Service was becoming intolerable. The outcome of all this was the Health and Safety at Work etc., Act 1974 which came into force in April 1975. Instead of the industry-specific, absolute but limited requirements which were the basis of the pre-1975 approach to workplace health and safety, the new Act required all employers to be accountable for the health and safety of their employees in all aspects of their activities.

THE EMPLOYER'S KNOWLEDGE OF THEIR OWN BUSINESS

Employers cannot discharge their duties under the Act unless they have an intimate knowledge of the dangers and hazards of their own undertaking. This statement is at the very heart of risk assessment which is now the common theme on which all health and safety management strategies have to be based. Employers need to know how to identify these hazards (physical, chemical, biological, ergonomic and behavioural), evaluate and control them. Employers who fail to do this are breaking the law and putting their employees and often the public at large at risk. It is as simple as that.

Many employers react with surprise if one suggests that they are not always fully aware of these hazards – especially hazards to health. Even after over a decade of the Control of Substances Hazardous to Health Regulations there are employers who regularly expose their workers to hazardous substances without any thought as to what effects these might have on their health. In many cases the employer has no idea how hazardous a substance may be; after all there are literally thousands of industrial compounds on the market these days, and new ones appear regularly as the catalogues of the relevant wholesalers will testify. The point is, the COSHH Regulations require employers to find out about such substances before their employees are exposed to them.

Case study 0.1

A company manufactures bespoke batching, bagging and other materials handling plant where the raw material, usually in powder or granular form, is automatically sorted, weighed and packed as required. Clients from many countries order plant to suit their specifications.

Before manufacture gets underway the customer is asked to provide a sample of the raw material the plant is to handle to ensure the finished plant performs to specification. The 'sample' amounts to a tonne or so of the raw material. On arrival at the factory these samples go straight into a sealed 'quarantine' warehouse. Small samples are then tested and analysed by the company laboratory for any dangerous ingredients before the material is released for test purposes. This procedure reduces the chance of exposure of the workers to any unknown chemical or biological hazard.

Employers and their employees need to know about toxicity, atmospheric pollution in the workplace, the effects on health, control measures and the acceptable standards and exposure levels appropriate to their activities. Employers particularly need to have sufficient knowledge of occupational hygiene to be aware of any potential dangers to health their technology, procedures and processes may generate. This book will help them to acquire that knowledge.

CHAPTER 1

Origins *of* occupational health *and* hygiene

A definition of hygiene is 'the maintenance of health and the prevention of disease'. Occupational health and hygiene applies this definition to the workplace.

The occupational hygienist is the person responsible for the maintenance of health and hygiene in the workplace, primarily by advising on exposures to substances, processes and environments generally which are likely to cause disease or ill health. Primitive hygiene measures were practised in ancient times to reduce the wastage of slaves. Although the relationship between work and disease is a fairly modern concept and only came into sharp focus with the coming of industrialisation, the ancient civilisations were aware of it.

EARLY CONCEPTS

These early concepts of health problems of the emerging industries were seen as traditional, that is, they were regarded as natural hazards of the job. There was widespread fatalism amongst industrial workers and their families at that time. Early deaths of Sheffield knife grinders, working long hours in airless basements (the original 'home workers') from lung diseases were taken for granted. Mill fever of the early textile mills – probably an early form of byssinosis – was just 'part of the job'. Until the scores of industry-specific regulations such as the Felt Hats Particulars Order 1903 (mercury poisoning), the Lead Smelting Regulations 1911 (lead poisoning), the Oil Cake Welfare Order 1929 (basic welfare facilities) began to identify specific industrial health risks, these risks were accepted by employees under the maxim of 'volenti non fit injuria' – the voluntary assumption of risk. Even after the Second World War 'dirt' money, 'dust' money and certainly 'danger' money was still available to workers willing to risk their health doing unhealthy jobs. At least, though, this practice indicated that employers were becoming aware of the existence of disease creating conditions in the workplace.

AMERICA TAKES THE LEAD

Only at the beginning of the twentieth century did occupational ill health begin to attract the medical response it sorely merited, but it was not until the 1930s that occupational hygienists started to appear and occupational hygiene blossomed into a recognised independent discipline. As with so many other cases it was the United States of America that provided the lead in this field, mainly due to pressure from workmen's compensation legislation. The rest of the world slowly followed, the UK running about ten years behind the US. Certainly the latter took the lead in establishing what are now termed 'Occupational Exposure Standards' (they were originally 'Threshold Limit Values') which we adopted several years later in most cases.

THE POST-WAR YEARS

The period after the Second World War saw considerable advances in general medicine as well as occupational medicine. The urgent need to provide effective treatment for servicemen suffering wounds and diseases caused the medical and health professions to develop on a significant scale. In particular, quantitative means to check actual workplace concentrations of contaminants (asbestos fibres, lead and other dusts and so on), which did not exist pre-war, were developed. This trend was accelerated, particularly in the case of asbestos, because asbestosis is a long-term process and it was only during the post-war years that many cases of the disease came to light.

Today, qualitative assessment has been elevated to an exact science with legal standards set for all industrial undertakings. Many of the more common standards are known as 'Occupational Exposure Standards', and are contained in HSE's Guidance Note EH40.

Case study 1.1

During the War years a Nottingham company assembled civilian gas masks on long production lines in the city. One of the components of the gas mask filter was a large 4" diameter white asbestos disc. On assembly this disc of flaky asbestos would be picked up by hand and placed inside the metal filter casing along with other components.

A few years ago a post-mortem on two sisters who had died in their eighties revealed asbestosis as the primary cause of death. The cause of the disease was a puzzle until an elderly friend remembered they had both been employed assembling gas masks during the War. It was then that they had contracted the disease.

THE CHRONIC NATURE OF OCCUPATIONAL ILL HEALTH

Work related ill health is often chronic in nature, that is, it develops over time and is not regarded by the media as sensational or even interesting. When the Robens Report was published in 1972 *The Times* newspaper commented, 'Not exciting, except by accident'. Occupational ill health hardly merited a mention yet thousands of cases existed at the time and thousands more were developing. This chronic – as opposed to acute – phenomenon seemed not to galvanise the reformers to any great extent in the early years of occupational health research, yet today over 30 per cent of *all* ill health is occupational and a quarter of this percentage is due to overexposure to chemicals.

HEALTH CARE FOR WORKERS

There are two dimensions to this subject:

Those in industry who are responsible

Have a good, up-to-date knowledge of work environment health hazards.
Often know little about workers' social/domestic backgrounds.

The worker's general practitioner

Working knowledge of domestic type illnessess but often knows little about worker's work environment. Should be capable of recognising embryonic problems/work symptoms. Surgery usually full of work related cases of sickness.

Initially there was little contact or cooperation between the two parties, but matters are slowly improving. By comparison, Europe's approach seems far more enlightened. In the years leading up to the 1974 Health and Safety at Work Act Britain tended to ignore recommendations from the International Labour Organisation to improve occupational health. It then created the Employment Medical Advisory Service (EMAS) alongside the Health and Safety Executive to make up for years of neglect.

It does seem rather ironic that British employers have never been legally obliged to provide other than very basic health services for employees, such as those laid down by the old Factories and Offices, Shops and Railway Premises Acts, yet the NHS spends millions on curing sick people and a pittance on preventative medicine.

THE EMPLOYMENT MEDICAL ADVISORY SERVICE

This service was set up in 1974 alongside the Health and Safety Executive. Its medical advisory staff have the same authority – powers of entry etc. – as HSE inspectors. Its primary function is to give advice on all aspects of occupational health and aims to promote awareness of health related matters in workplaces. It is staffed by doctors and nurses – many on a retained basis – with specialist qualifications in occupational health. They advise employers on needs of disabled employees as well as identifying their general occupational health needs. They also carry out investigations of suspected or confirmed cases of occupational diseases and will look into disputes between unions and management where a general difference of opinion on a health matter exists.

Case study 1.2

A large egg packing company started to receive complaints from female operatives who worked in egg candling booths. These are darkened booths where eggs pass over a ground glass plate illuminated from underneath. The candler sits on a chair and spots eggs with defects, the light passing through the shell.

The Union at the time adopted a militant stance claiming that the candling process was harmful to the women's eyes as they were complaining of strain, fatigue and other physical symptoms. The company enlisted the services of the local Employment Medical Adviser who pronounced, after investigation, that the problem lay with the women, all of whom should have been wearing glasses, and that the process was perfectly safe. The Union accepted the Adviser's report without a murmur. The exercise cost the company nothing.

If an occupational health problem exists or develops on your premises or among your employees use the Employment Medical Advisory Service. It can be contacted at your HSE Area Office.

SUMMARY

Occupational health care in the UK has developed in fits and starts and on average has lagged behind both Europe and America. The pervading theme through much of the time occupational health knowledge was developing was the reluctance of 'those in charge' to accept that a particular substance, process, by-product, industry and so on was the cause of a specific disease, illness or medical condition when virtually everything pointed to it being so. The insistence that 'there was no proof' that working in a particular industry with a specific chemical caused a specific illness when only workers in that environment contracted that illness makes one wonder just what proof was required.

Unfortunately, there is a political side to occupational health. Where an industry produces revenue as well as ill health the prevarication regarding 'proof' can be long and devious. But in an industry such as the asbestos industry, which does not produce revenue, 'proof' and the clamour to 'shut it down' are readily forthcoming.

Fortunately, the coming of the Health and Safety at Work Act heralded major improvements in occupational health research and management. Sadly, the high hopes and aspirations of EMAS have been severely curtailed due to lack of funding over the years but it does act when circumstances warrant this.

The problem that persists is that of cleanliness, hygiene and housekeeping in many of today's businesses, and management's attitude towards these matters. In fairness, though, the contribution of many workforces towards personal, as well as occupational, hygiene leaves something to be desired. The condition of a company's washrooms is always a good, initial indicator of the general occupational hygiene standards maintained by that company.

CHAPTER 2

Occupational health risks

Risks to the health of people at work can be classified under four headings: physical, chemical, biological and ergonomic.

The last named category can be classified under physical risks but in view of the high profile it enjoys from the DSE Regulations it is useful to classify ergonomics separately. Radiation is classified as a physical risk.

PHYSICAL RISKS

Examples of physical risks and their effects are as follows:

Noise – noise-induced hearing loss. Progressive and permanent. Tinnitus usually develops which is also permanent. Stress, fatigue, distraction.

Vibration – hand and arm affected by vibration induced 'white finger'. Blanching of the fingers/thumb accompanied by considerable pain.

Barometric pressure – decompression sickness (the 'bends').

Heat – heat cataract, heat stroke, heat exhaustion, dehydration, heat cramps.

Light – glare, eyestrain, wrong type of lighting, miner's nystagmus, headaches, fatigue.

Ionising radiation – chemical and cell structure changes in the body. Biological effects. Cancer, skin diseases, radiation sickness.

Non-ionising radiation – extreme heating of material or body parts exposed (e.g. microwaves).

CHEMICAL RISKS

Exposure to chemicals at work can result in various forms of poisoning, diseases and damage to the central nervous system. Some examples include:

Acids and alkalis – burns to the skin and flesh, non-infective dermatitis.

Metals (and compounds) – poisoning and organ damage e.g. lead, mercury, chromium.

Non-metals – poisoning and organ damage e.g. arsenic, vinyl chloride, phosphorus.

Gases – poisoning, organ and nerve damage. Arsine, carbon monoxide, chlorine.

Organic compounds – notorious for causing occupational cancers.

BIOLOGICAL RISKS

These risks are created by exposure to various forms of bacteria, viruses, spores and dusts. They are usually classified as human-borne risks (including blood-borne) such as viral hepatitis and AIDS; vegetation-borne such as aspergillosis ('farmer's lung' from mouldy hay); or animal-borne such as glanders, anthrax, brucellosis and even foot and mouth disease and possibly variant Creutzfeld-Jakob disease. Weil's disease (from rats' urine) and psittacosis (from birds, e.g. game and parrots) are other examples of diseases transmissible from animals to humans.

ERGONOMIC RISKS

These are normally of a musculoskeletal nature and develop in situations where abnormal posture along with frequent repetitive movements place excessive stress and strain on ligaments, tendons, muscles and the nerve endings. Poor workstation layout – especially in the case of display screen equipment users – contributes to a whole group of work related upper limb disorders which are caused by repetitive strain injuries (RSI). Some forms of RSI are given below:

Tenosynovitis – inflammation of the tendon sheath which provides lubricant for the tendon moving inside it.

Carpal tunnel syndrome – all the nerves and tendons of the hand pass through a sheath near the wrist, rather like cables in a wiring harness, called the carpal tunnel. This tunnel becomes inflamed with overuse of the tendons passing through it, creating a very painful condition.

Tendinitis – inflammation of the tendons, particularly the ones in the hand and fingers.

Epicondylitis – inflammation of an area where a muscle joins a bone.

Peritendinitis – inflammation of an area where a tendon joins a muscle.

Cramp, writer's cramp – the involuntary, slow, forcible and very painful contraction of a muscle brought on by overexertion, fatigue, or poisons affecting the nerves controlling the muscles.

For many years British medical circles did not recognise work related upper limb disorders as a medical condition although the International Labour Organisation recognised RSI as an occupational disease in 1960. By the time the UK *did* recognise RSI many workers had already developed symptoms and some quite considerable damage settlements were forming in the pipeline. In 1994 an Inland Revenue employee was awarded £79,000 after contracting RSI during her 14-year employment.

CHAPTER 3

Physical health risks

In this chapter an explanation of some common physical health risks will be made and suggestions on practical measures for the control of employee exposure provided.

NOISE

It does not matter whether this is defined as 'noise' or 'sound', if the amount of it is excessive it can cause occupational deafness or 'noise induced hearing loss'. A wide range of sounds can cause this deafness and it is the frequency, duration of exposure and intensity of the sound that causes the damage to the delicate structure of the inner ear. Contrary to popular belief, removal of an employee who has been overexposed to noise to a quieter work environment will *not* help to repair the hearing loss. This is progressive from the time it starts. As the person gets older the natural hearing deteriorates anyway and this is 'added' to the occupational hearing loss. In addition, a really dreadful condition known as tinnitus can develop. This is permanent noise in the ears, night and day, without respite, and can take a variety of forms from a swishing sound to that resembling a pneumatic riveting hammer.

Excessive noise may affect the hearing in three ways:

1 **Temporary threshold shift** – short term loss of hearing acuity. The hearing recovers after a time depending on several factors relating to exposure time, noise intensity and individual susceptibility. Visits to very loud discos can produce this effect.

2 **Permanent threshold shift** – in this case there is some recovery but it is never complete, hence the name of the condition. Again, the degree of permanence is dependent on exposure time, noise intensity and individual susceptibility.

3 **Acoustic trauma** – this condition can result from percussive noise of high intensity and short duration such as a gunshot, explosion or generally a very loud bang. A blow on the head near the ear as well as percussive noises can rupture the eardrum and damage or dislocate the delicate components of the hearing mechanism.

Practical control measures

Noise control measures are often difficult and expensive but it pays to look closely at the way noise is being generated and transmitted in the first place. Often, where metal to metal or metal to concrete contact is being made, simple cushioning arrangements with wood, rubber, cloth and other sound absorbent materials can be effective and cheap.

Where machinery is involved acoustic enclosures or isolation can be considered, e.g. site compressors outside a building. If vibration is a noise source, damping with anti-vibration mountings may be effective. Fan noise can be reduced by altering the pitch of the blades; rotating components can be dynamically balanced and so on. Noise pathways can be obstructed by noise-insulating or noise-absorbing barriers. Finally – and as a *last* resort – hearing protectors should be worn. They must be worn the whole time a worker is exposed; removing protectors for even short periods (three or four minutes) will nullify the protection for the whole working day.

VIBRATION

Generally referred to now by the HSE as 'hand-arm vibration' the main occupational disease related to exposure to vibration is 'vibration white finger'. However, regular exposure to high levels of vibration can lead to permanent injuries to the fingers, hands and arms, collectively known as hand–arm vibration syndrome. The injuries can apply to the blood circulatory system. For example:

• vibration white finger – where the fingers become white, numb and very painful
• the sensory nerves – you cannot pick things up and lose your sense of touch and temperature
• the muscles of the wrist, hand and fingers – loss of grip
• bones and joints – pain, tingling and numbness in the hands, wrists and arms which often disturbs sleep.

Any power tool or equipment with a vibratory action or component where vibration is transmitted from the work equipment or the process into the workers' hands or arms can contribute to the development of hand–arm vibration syndrome.

It should be stressed that occasional exposure to a vibrating tool or process involving vibration is most unlikely to cause injury.

Raynaud's syndrome

This is a disease which manifests similar symptoms to vibration white finger but is not associated in any way with exposure to vibration. It is thought to be hereditary and results primarily from diseases of the connective tissue. People suffering from Raynaud's syndrome should not use vibratory equipment.

Practical control measures

There are a number of recommendations relating to equipment which have recently been drawn up, and where the purchase of new equipment is concerned, manufacturers have to abide by strict standards on vibration performance. Information on vibration levels has to be provided and equipment with the lowest vibration should be acquired. Vibration is measured in units of metres per second squared (m/s^2).

Apart from the equipment side, workers using vibratory equipment should maintain a good blood circulation by keeping warm and dry, reducing smoking and exercising and massaging the fingers during work breaks.

The HSE produce some excellent material on HAV for employers. A booklet, *Health Risks from Hand Arm Vibration INDG175 (rev 1)* is available free of charge from HSE Books, along with a very well written pocket card for employees. *Hand–Arm Vibration Syndrome – Pocket Card for Employees INDG296P* is available in moderately priced packs of 25. Employers in scope are well advised to get hold of these publications.

BAROMETRIC PRESSURE

Where people are working in environments where the pressure is different, usually greater, to atmospheric pressure (1 bar = 10^5 Newtons per square metre), entering and leaving those environments must be done with great care. If the transit period is too rapid bubbles of air may come out of solution in the bloodstream and block small blood vessels causing severe muscle pains (the 'bends'), tingling and choking sensations and sometimes paralysis or coma.

Practical control measures

This is a highly specialised subject requiring a great deal of expertise and control. Basically, though, if a person shows symptoms of the 'bends' the correct treatment is *recompression*, i.e. put the employee back where they were in the different pressure environment and then provide them with slow decompression.

Construction of the Channel Tunnel had to be carried out with men working at pressures higher than atmospheric pressure. There were some cases of decompression sickness during that time due to failure to adhere strictly to decompression procedures.

TEMPERATURE

The UK has in the past been regarded as a fairly cold country and its buildings reflect that fact. Although there are still some work environments where occupational conditions associated with excessive heat exist, problems with temperature control of office and similar working accommodation are emerging due to the hot summers of late. Employees are, on occasions, being sent home because the office temperature is far too high for comfort.

Normally the human body's thermoregulatory system varies the body's temperature to cope with wide variations in the external temperature. However, if these external variations are too great or the body's regulation fails then *heat stroke* can occur if the body temperature reaches 40.5° C (105 – 108° F) or higher. A person exposed to excessive heat may fall unconscious due to suppression of sweat secretion. If the body cannot sweat it cannot lose heat. Should this occur the body temperature must be brought down to 40° C as quickly as possible otherwise there is a possibility of damage to the central nervous system.

Workers involved in furnace work such as glass blowing, can contract *heat cataracts* of the eyes caused by excessive exposure to radiant heat and microwaves. The correct eye protection must be worn at all times in such situations.

Where excessive sweating occurs under hot conditions *heat exhaustion* can cause abdominal cramp, an increased pulse rate and dizziness with pains spreading from the calves to the arms. This is due primarily to body salt being lost through sweat. A saline drink will ease the condition but for some people such a drink, or the swallowing of salt tablets, will induce vomiting.

In hot weather employees should drink plenty of water, otherwise *dehydration* can set in very quickly. This is indicated by yellowing of the urine accompanied by a stronger than usual odour.

Many people are keen on getting a tan and strip off the moment the sun shines if working outdoors. The fact is that the danger to the skin from *sunburn* is ignored by most people. Although the HSE are publicising these dangers – principally from unfiltered ultra-violet radiation – employees, mostly in the construction industry, are ignoring these warnings. Apart from the pain that sunburn brings, the possibility of skin cancer developing is very real.

Practical control measures

It is important to have adequate supplies of drinking water available in all work environments where heat is a problem – in fact it is a legal requirement. It is also necessary to get employees to drink it. There is a strange reluctance to drink water in the UK. It has to be requested in most restaurants and only comparatively recently have bottled water supplies with chillers started to make a welcome appearance in offices and other workplaces. Medical experts recommend that everyone drinks three pints of water a day for health reasons regardless of their occupation.

Outdoor workers should 'cover up' or 'keep their shirts on' as per the exhortations of the HSE literature. If exposure of skin to the sun is insisted upon then an appropriate filter cream must be applied.

Loose, lightweight clothing should be worn where excessive heat exists. Circulation of air around the body helps to keep it at the right temperature.

Where employees work in conditions of extremely low temperatures such as cold rooms, refrigerated warehouses and similar there is a wide range of insulated clothing available. Other than that, the law requires warm refuges to be provided and that continuous work in the cold environment does not amount to unreasonable periods without breaks.

LIGHTING

Poor lighting of a work area or lighting of the wrong type can cause accidents but more commonly causes eyestrain, visual fatigue, vertigo and headaches. Good lighting is an exact science but many managers are unaware of this. Lighting can be too bright but in the majority of poor lighting cases it is at too low a level. The wrong type of lighting does have a psychological effect on people – the famous Hawthorne experiments proved this – and it has an effect on production and absenteeism.

Practical control measures

If you are planning new premises or upgrading existing ones it is really worth while engaging the services of a firm of genuine, recommended, lighting consultants. Ask the employees who are working indoors what they think of the lighting. Many old 'popular pack' fittings with 'cool daylight' fluorescent tubes of the 1950s, 60s and 70s are well past their expiry date and should be replaced. Modern fittings are brighter, more economical and have a far superior colour rendering.

RADIATION – IONISING

Radiation is defined as 'the emission and propagation of energy through space in the form of electromagnetic waves or subatomic particles'. This propagation can range from harmless radio broadcasting waves to dangerous gamma rays. Radiation, an emotive term often misunderstood, now covers all

11

Case study 3.1

The Hawthorne Experiments were a series of studies of workers' reactions to changes in working conditions and practices carried out in the 1920s by Elton Mayo and others at the Western Electric Company's plant at Hawthorne, Chicago, Illinois. A group of female telephone equipment assemblers was put to work in a special room and their reactions to the changes observed. One of the major changes involved the type and level of the lighting. As the lighting in the room was improved, productivity, morale and motivation also improved. Working extra hard because of feelings of participating in something new and special and being consulted about changes in working conditions became known as the 'Hawthorne effect'.

electromagnetic waves such as light, radio and television waves, X-rays and the particles emitted by radioactive materials as they disintegrate or decay to reach a non-radioactive state. These particles and the more energetic electromagnetic waves produce electrically charged particles (called 'ions') in the materials they strike.

The basic structure of all matter, including human tissue, is the atom which consists of a nucleus made up of positively charged protons with an equal number of negatively charged electrons which orbit the nucleus. Such an atom is said to be balanced. However, there are other forms of the same atom which have *neutrons,* as well as protons, in the nucleus. These atoms (of the same matter) are called *isotopes* and the number of neutrons equals the number of

protons which is the same as the number of orbiting electrons. There are other isotopes, though, which have additional neutrons in their nucleus and these neutrons cause the atom to be unstable. These extra neutrons can spontaneously change into protons and electrons and the latter are emitted from the atom's nucleus at very high speed. Isotopes possessing these properties are known as *radioisotopes* and are thus *radioactive.*

It will be seen that this radioactivity, or radiation, will ionise materials exposed to it, and where human tissue is concerned, chemical changes leading to damage to such tissue will occur. This affects the behaviour of the body's cells by either damaging the DNA structures present in every cell or killing the cells completely. Providing the number of cells killed is not too high the body replaces these by natural means. It is the damaged cells which affect the individual exposed (the somatic effects) and possibly subsequent generations (hereditary effects).

Types of ionising radiation
Alpha particles
Consist of protons and neutrons. Easily stopped and do not penetrate the skin. Only hazardous if alpha emitters are swallowed or inhaled into the body – or enter through a break in the skin. Alpha radiation can be stopped by a sheet of paper.

Beta particles
Consist of electrons. Have greater penetrating powers than alpha particles but are stopped by glass, water or sheet metal. Again, hazardous if taken into the body.

Gamma radiation and X-rays
This is electromagnetic energy without mass or charge travelling in wave form at the speed of light. Note that X-rays have much less energy than gamma rays hence their use for medical purposes. They are stopped by sufficient thicknesses of lead and concrete to absorb the energy involved.

Neutron radiation
Neutron particles which are very penetrating but can be stopped by thick layers of concrete or water.

RADIATION – NON-IONISING
Non-ionising radiation does not have the effect of altering the make-up of living cells or causing chemical or biological changes to the body. However, it is still potentially dangerous in another way. Non-ionising radiation is usually absorbed by the body cell structure causing it to heat up in the same way that a microwave oven operates.

Types of non-ionising radiation
Lasers
The term 'laser' is an acronym which has now entered the English language as a word in its own right. It stands for 'light amplification by stimulated emission of radiation' and a laser is a very high energy beam of light. Industrial lasers can cut through quite thick metal by burning and can produce far more intricate and delicate patterns than can be achieved by oxy-acetylene burning, for example. Carelessness with such lasers can result in a hole through the finger or hand very quickly indeed. Other lasers, such as laser pens, are very dangerous as far as the eyes are concerned. Lasers also emit infra-red radiation (see below under 'Infra-red radiation').

Ultra-violet radiation
In the electromagnetic spectrum ultra-violet (UV) radiation is the next region to gamma radiation. It has longer wavelengths than gamma radiation which renders it non-ionising. Two common sources of ultra-violet radiation are the sun and electric arc welding. Injuries to the eye that can be caused by UV radiation include keratitis (damage to the cornea) and photochemical cataract. Effects of UV radiation on the skin include erythema (severe reddening of the skin – 'sunburn'), skin ageing and skin cancer.

Infra-red radiation
This type of radiation is associated with radiant heat such as emitted by the radiant bars on an electric fire, but all hot objects emit infra-red radiation. Again, it is the eyes which are

vulnerable and long term exposure to even low doses of infra-red radiation can burn the lens and cause heat cataracts.

Microwaves and radiofrequencies

Radiation at this level is a prime source of internal heating of the body organs. It is produced by high frequency transmitters such as those used by radar and certain communications equipment. Heat sealers, RF heating coils and microwave ovens generate microwaves, the last mentioned item provides a good example of the heating effects of non-ionising radiation.

Practical control strategies
Ionising radiation

It is obvious that ionising radiation is dangerous and care must be exercised to ensure employees are not overexposed to it. The term 'overexposed' is used because people working with ionising radiation do receive doses from time to time. An organisation known as the International Commission on Radiological Protection has published masses of data on the effects of ionising radiation on the body and has set recommended dose limits with the aim of limiting the risks to individuals from cancer and hereditary effects to acceptable levels and of completely preventing the occurrence of other types of damage to tissue of the body. Updated information can be obtained from the HSE or the National Radiological Protection Board (NRPB) at Didcot, Oxfordshire.

In lay person's terms protection is based on three factors; *time,* that is the length of the period of exposure, *distance,* or how far away you are from the source of radiation, and *shielding,* or what type of barriers can be erected between the employee and the source

and type of radiation. A study and understanding of the Ionising Radiations Regulations 1999 is essential if you are involved with any sources of ionising radiation. The Regulations impose a hierarchy of measures for controlling exposure which include shielding, ventilation, containment and minimising of contamination. Continuous monitoring of personal dose levels must be made, and warning signs, designated controlled areas and prevention of leakage, must be strictly observed.

Non-ionising radiation

Most of the precautions for ionising radiation apply equally to non-ionising radiation. It should be stressed that the eyes are as vulnerable as any organs in the human body and should be well shielded from lasers, UV and infra-red rays. In some ways non- ionising radiation can inflict more damage on the body due to an attitude of indifference and ignorance towards it. Where non-ionising radiation sources exist, those responsible must ensure that employees are provided with information about the hazards such sources present and what means of protection should be used.

DISPLAY SCREEN EQUIPMENT

Since the use of personal computers in particular and display screen equipment in general has become virtually universal (are there any employees sitting at desks still without one?) more and more research has gone into stress, all kinds of ill health and a group of musculoskeletal disorders known as 'work related upper limb disorders'. The science of ergonomics, for years confined to what was commonly referred to as 'the man-machine interface', has assumed long deserved respectability and recognition amongst many

```
Case study 3.2

In 1979 BNFL allowed a leak of more than 10 cubic
metres of radioactive liquid into the ground
surrounding its Windscale (now Sellafield) plant
in Cumbria. The leakage was subject to an official
inquiry by the Nuclear Installations Inspectorate
who found the cause to be a combination of poor
plant design, poor management control and a failure
to establish Operating Limits for radioactive
material as required by the Company's licence. An
interesting finding of the Inquiry was that the
level of radioactive liquid in a sump vessel rose
to a dangerous level and was not detected. The
reason for this was because the sump content gauge
was taken as indicating a low reading when in fact
its pointer was on the second circuit of its circular
scale!

(Source: HSE publication ISBN 07176 0061 0, 1980)
```

employers who previously held dismissive views on its principles. One reason or this change in attitude is the quite large damages settlements awarded to employees developing, or who have developed, upper limb disorders. Yet even the threat of incurring such settlements against them seems not to alert some employers to the potential ill health problems of using display screen equipment or the legal requirement to assess the risks of their employees' workstations. Many public sector employers still seem indifferent to the Health and Safety (Display Screen Equipment) Regulations 1992 whilst pursuing far less important legal requirements with some fervour.

Many of the ill health problems associated with working with display screen equipment for long periods – even with regular short breaks – are listed on page 7. General fatigue, both physical and mental, is most often associated with such work. Eyestrain and eye fatigue are also quite common. The root cause of most ill health conditions stems from the fact that many thousands of employees simply had a base unit, display unit and keyboard dumped on their already inadequate, standard, rectangular desk, with fixed chair, shown how to connect it up and then told to 'get on with it'. There were two positions for the screen; either at one corner of the desk which necessitated twisting the neck when keying with the keyboard directly in front of the user or in the centre of the desk which brought the keyboard and user too close to the screen. Either layout is manifestly unsatisfactory but in so many cases there is no alternative. Fortunately the more enlightened employers are now providing wrap-around desks which are specifically designed to accommodate display screen equipment.

The basic requirements for a DSE workstation are contained in The Schedule and Annexes to the 1992 DSE Regulations. They are fairly detailed but boil down to straightforward common sense. The requirements can be checked by using a simple checklist, discussing the workstation layout with the occupants and using one's eyes in carrying out a risk assessment. Above all, DSE workers should be instructed to tell their managers about any developing discomfort. The fact that this instruction has been given should be recorded. When an employee does complain about any discomfort the complaint *should be acted on* and the results recorded.

```
Case study 3.3

A Northern Local Authority faces having to pay up
to £250,000 compensation to an ex-employee who
developed back and neck pains soon after display
screen equipment was installed in her workspace. No
risk assessment was carried out prior to this
installation and as the individual was only 5'2" in
height her feet did not touch the floor. No footrest
was provided and the equipment layout, already in a
cramped area, necessitated constant twisting from
her desk to her screen. After an initial period off
work the employee returned to find an increased
workload awaited her and she was not allowed to
take the regular breaks required by the Regulations.
Representations to her employers fell on deaf ears
and the employee developed cervical and lumbar
spondylosis (immobility of the vertebral joints).
Meanwhile the employer, who was in gross breach of
the Display Screen Equipment Regulations, was
actively prosecuting a market trader for selling
fruit in pounds instead of kilos!
```

There are still some lingering myths about eye damage and radiation associated with display screens. There is no danger whatsoever of damage to eyesight but eyestrain and eye fatigue can occur. This is often a result of poor work schedules, poor workstation layout and insufficient breaks. It is worth quoting Professor John Marshall of the Moorfields Eye Hospital, an acknowledged expert on visual display units:

'There are no radiation hazards from visual display units. In consequence there are no eye health hazards from visual display units. There are, however, problems of eye comfort, but these usually relate to working area design and proper setting up of the VDU. Comfort, however, should not be confused with health'.

Work at display screen equipment will not cause epilepsy or harm to pregnant women from radiation. It is significant that employees harbouring concerns about 'radiation risks' from office VDUs apparently remain untroubled by sitting at home in front of television screens much larger than computer display screens!

Practical control strategies

Research has shown that the majority, if not all, of the symptoms described by VDU users are a reflection of bodily fatigue. In many cases the symptoms experienced arise from a number of causes with the result that it becomes difficult to attribute individual symptoms to a particular aspect of the workstation. It is, therefore, advisable to divide the symptoms into three categories:

- firstly, those symptoms related to the **visual system**
- secondly, those related to the **working posture**
- thirdly, those related to the nature and organisation of the **work itself.**

In assessing the risks inherent in DSE work and in the training of risk assessors, factors arising in these three discrete categories must be thoroughly addressed. This is neither a complex nor a difficult course of action. Problems arising in the first two categories can be largely overcome by the application of straightforward ergonomic principles in the design, selection and installation of VDUs and the associated design of the user's workstation, including the physical environment.

As regards the third category, attention must be given to the design of the operator's task since this is likely to be at least as important as the first two categories in determining the acceptability of the workstation as a whole.

As mentioned above quite simple and straightforward guidance towards achieving a satisfactory standard for working with VDUs in the above three categories is provided in the Schedule and Appendices to the Health and Safety (Display Screen Equipment) Regulations 1992.

SUMMARY

This chapter has looked at the more common, everyday, physical risks to health at work from a down to earth, lay person's perspective and suggested basic, practical control measures to reduce their effect on employees' health. Many of these measures are based on the author's years of experience with such risks at work and contact with sufferers of, for example, hand-arm vibration, noise-induced hearing loss, laser burns and musculoskeletal disorders from DSE work. Where myths exist, e.g. in the case of radiation with display screens, these have been debunked. The genuine dangers of temperature, radiation, excessive noise and so on have been put into their correct perspective and mini-case histories – all in the public domain – provided where these may be of interest to the reader.

Physical **health** *hazards are, of course, closely associated with physical hazards themselves and these are legion. Health hazards of the construction industry would be the subject of a separate volume. But whilst this book was being written an incident which nearly resulted in the death of a workman was reported in a television programme (see case study 3.4).*

Case study 3.4

A workman was laying pipes in a deep, narrow trench dug into a road. The trench was about 18″ wide and the sides were not supported in any way. It was over 6′ deep.

A section of one wall of the trench where the pipelayer was working gave way, pinning his legs and waist to the opposite wall. The fire brigade worked frantically to free him as the blood circulation to his legs and feet began to fail.

After a long and difficult struggle the workman was freed and eventually recovered. It was initially assumed that he might lose both his legs above the knee due to the medical problems he experienced concerning blood circulation rather than his physical injuries.

On a different note it is difficult to comprehend how the workman could have entered a deep, narrow trench to lay pipes without adequate support members securely fixed against the soil faces. The correct support is a basic tenet for trench work in the construction industry.

CHAPTER 4

Chemical health risks

The Health and Safety at Work Etc., Act 1974 states, inter alia, *that an employer must ensure, so far as is reasonably practicable, safety and the absence of risks to health in the 'use, handling, storage and transport of articles and substances' at work.*

This simple, unpretentious statement is a clever piece of legal drafting in that there are

- no activities at work whatsoever which do not fall into one of the categories 'use', 'handling', 'storage' or 'transport'; and
- everything used at work is either an 'article' or a 'substance'.

All items at work which can create a chemical risk to health are, therefore, substances and these can be in one of the following forms:

- **Gases** – for example, chlorine, phosgene, carbon monoxide, hydrogen cyanide and ammonia
- **Vapours** – particularly solvent vapours, for example, trichloroethane, xylene, toluene and volatile flammable liquids
- **Liquids** – flammable liquids such as petrol, acetone, methylated spirits, paraffin, oil and liquefied petroleum gases
- Other forms of non-particulate pollution
- **Dusts, powders** – including fibres such as asbestos, mineral dusts causing mesothelioma and fibrosis, cotton and textile

dust producing byssinosis and siderosis from iron dust
- **Mists** – various aerosols which generate minute droplets of liquid in a sprayed mist form, for example, paint spray, WD40 freeing agent and others
- **Fumes** – these usually contain minute metallic particles such as welding fumes and vehicle exhaust emissions
- **Smoke** – unburnt particles of products of combustion, steam, fumes, vapours
- Other forms of particulate pollution.

TOXICITY AND HAZARD

The danger to human beings from hazardous chemicals stems from their toxicity. If a chemical possesses toxic properties these will be present in whatever form the chemical takes. The overall goal of the industrial or occupational hygienist is to protect workers from the health hazards of their working environment. In this context and in today's high technology work environments there will always be toxic and potentially hazardous material to be processed or handled. This handling will incur physical stressors such as noise, heat and perhaps radiation as well. In devising appropriate measures to control exposure to substances hazardous to health the occupational hygienist, or other responsible person – particularly in the SME – must be sensitive to the cost effectiveness of all proposed measures to control alleged exposures. Before such measures are considered, however, it is necessary to understand and clarify the terms *toxicity* and *hazard.*

Toxicity

The toxicity of a material or substance is a natural property of that material which is fixed and cannot be altered. It can be defined as its effects upon a living organism – i.e., an employee. However, *individual susceptibility* to a toxic substance can vary and does so with age, physiology, sex, state of health, diet and so on, rather like colds and flu. Some people never contract these conditions; others seem never to be free from them.

Hazard

All substances are to some degree toxic but those of concern in occupational health matters are those which are or could be damaging to workers' health in amounts which are ordinarily, or likely to be, present in the workplace. This is when the toxic material becomes a *hazard.* In the previous chapter the hazards of radiation are explained. There is natural radiation all around us – from granite, the sun, the ground – but this is at a level where it does not pose a hazard. It is not generally known that companies which manufacture food colourings, flavours and fragrances (yes, *fragrances*) use trace quantities of poisons in their production to enhance the end product. But the quantities are so minute the toxic ingredients are not a hazard to the human body.

Measuring toxicity

The toxicity of a substance cannot be altered but different substances have differing degrees of toxicity. Data on this phenomenon is available in different forms but the most common method of measuring toxicity is to use the LD_{50} or the lethal dose of a material which will kill 50% of a population of laboratory animals, usually rats. This can be a single dose measured in milligrams per kilogram of body weight of the test animal to determine the *acute* toxicity of the substance or by a long term programme of repeated doses to determine the *chronic* effects of the substance.

There are other measures of toxicity. For example, LD = lethal dose; LC = lethal concentration; LC_{50} = lethal concentration for 50% of a specified population; LC_{LO} = lowest published *lethal* concentration and MLD = minimum lethal dose.

It is, therefore, quite possible for the lay person to make broad comparisons between the level of toxicity of different dangerous substances by studying the suppliers' data sheets (required to be provided under the Control of Substances Hazardous to Health Regulations 1999).

A simple example will make this clear. Imagine a *hypothetical* situation in which an employer had the choice between two chemicals for a particular process. One was arsenic in black crystalline form and the alternative was mercuric chloride. Arsenic is quite well known to be poisonous and appears to have been popular amongst the Victorians for despatching unwanted souls. A manager, knowing of the notoriety of arsenic as a poison might demur from using it in the process under discussion and opt for mercuric chloride instead. However, on checking the toxicity of these two substances these are revealed as follows:

Arsenic – subcutaneous* LD_{LO} (rat) = 25mg/kg of rat body weight (The lowest published lethal dose)

Mercuric chloride – subcutaneous LD_{LO}(rat) = 14 mg/kg of rat body weight (The lowest published lethal dose)

It will be seen, therefore, that in theory it takes less mercuric chloride to kill rats than it does arsenic, and while both substances are deadly poisons, arsenic is slightly safer.

* 'Subcutaneous' means beneath the skin e.g. via an injection or via a break in the skin or a cut.

EFFECTS OF TOXIC SUBSTANCES ON ORGANS AND TISSUE

There are no uniform effects of toxic substances on individual bodies as physiological differences between individuals – already mentioned above – amongst other things, lead to different reactions. A simple example is smoking. Nicotine has a toxicity rating of oral LD_{50} (rat) = 53mg/kg, which is high, but some lifelong smokers long outlive non-smokers. Others die of cancer in their early years. As far as exposure to toxic substances at work is concerned, control of exposure should be based on an intelligent interpretation of information from HSE literature and if a large number of chemicals are involved, reference to a directory of dangerous properties of industrial chemicals. Such directories are compiled, for example, by N Irving Sax and published by Van Nostrand Reinhold. There are other such publications for easy reference.

The **brain** is very susceptible to lead, mercury and heavy metal poisoning. Solvents can cause quite severe damage. The Emperor Nero, allegedly fiddling while Rome burned, is said to have been suffering from acute lead poisoning. This was occasioned by drinking acidic wines from lead drinking vessels. Certainly his behaviour was quite abnormal.

In the case of the Mad Hatter from Lewis Carroll's *Alice in Wonderland* it is well known that salts of mercury were used extensively in felt hat manufacture. Such hat manufacturers were exposed to these salts for long periods thus sustaining severe brain damage. The Felt Hats Particulars Order 1903 was promulgated to protect such workers and remained in force until repealed by the Control of Substances Hazardous to Health Regulations 1989.

Nasal passages. Areas such as the nasal passages and other areas with mucous membranes are vulnerable to nickel and chromium compounds which cause severe inflammation, ulcers and even cancer.

Lungs. Fuming acids e.g. nitric acid; chlorine, ammonia, oxides of nitrogen, solvents such as trichloroethane can permanently damage the lungs. The wide range of respirable, toxic dusts such as asbestos, silica and coal dust cause permanent scarring of the lung lining, rendering it unable to perform its job of passing oxygen into the bloodstream and removing carbon dioxide from it during breathing.

Lungs and skin. Isocyanates are particularly dangerous to the lungs and skin. Toluene di-isocyanate can damage the lungs irreparably in a very short time.

Case study 4.1

Some years ago a 36-year-old man was filling the space between the inner and outer casings of refrigerators with liquid foam toluene di-isocyanate. TDI is the basis of many foam products and was widely used for refrigeration insulation. It is highly toxic whilst in its hot, liquid foam state but is completely inert when cold and set.

During the filling process the pressurised TDI delivery line sprang a leak and molten foam TDI started to spurt out onto the factory floor. The line was switched off and a foreman told the man to get under the filling line and clean up the foam spillage with a paint scraper before it hardened. The smell was so strong the foreman fetched a paint covered respirator from the paint shop for him to wear. It was the wrong type and leaked around the edges. The man worked cleaning up the mess for an hour and a half. After a few days he became breathless and quite distressed. He saw the factory doctor who diagnosed asthma.

In fact the TDI vapour had destroyed over 30% of his serviceable lung lining in the short time he was scraping up the spillage. The company* knew of the dangers of TDI but did not pass this information down the line. The works medical officer and the safety officer were more concerned with covering up – the company initially denied the accident had happened. Eventually the man was awarded £4,000 for a ruined life when the company finally admitted liability.

* Frigidaire, a Division of General Motors, factory in North London, 1973.

The **liver** is particularly affected by chlorinated hydrocarbons such as trichloroethane, ethylene dichloride and perchloroethylene – chemicals used for degreasing and defatting. Many chemicals, e.g. ethylene chlorohydrin and dioxane, cause irreparable damage to the liver.

Kidneys. Substances affecting the liver invariably have a similar damaging effect on the kidneys. Exposure to such substances can cause acute nephritis or inflammation of the kidneys with haemorrhaging.

Bladder. Many toxic substances which were once associated with rubber and cable manufacture and dyeing were thought to cause bladder cancer. These are now either prohibited e.g. benzidine, or controlled, e.g. 2-naphthylamine.

Central nervous system. Mercury, cadmium, compounds of heavy metals, solvents, lead and certain toxic gases attack and damage the CNS, which is very susceptible to a wide range of substances.

Skin. Particularly sensitive to defatting agents such as detergents, solvents and chlorinated hydrocarbons. Lubricating and mineral oils and used engine oils are particularly injurious to the skin, quickly causing the onset of industrial dermatitis.

Bone marrow can be destroyed by benzene (now a prohibited substance) and vinyl chloride monomer. Hydrofluoric acid, if spilled on the skin, will penetrate skin, tissue and bone to attack the marrow.

It is important to remember that *all* toxic substances entering the body enter the bloodstream and are thus carried systemically to all parts of the body.

The body's defences against toxic substances

However, it is not all doom and gloom. Notwithstanding the effects of toxic substances on the body and its organs outlined above, the natural defences of both the body and individual organs are quite remarkable. Each route of entry into the body i.e. the *skin,* via absorption; the *lungs,* via respiration and *ingestion,* via the mouth, has a special membrane wall which protects the body from absorption of some classes of chemicals. The skin, for example, is nearly impervious to hydroxy, carboxy and ionised molecules, while hydrocarbons, esters and fats pass through with relative ease. Some compounds have greater mobility within the body that others. Benzidine placed on the skin can be detected in the urine within 20-30 minutes whereas benzidine hydrochloride and benzidine sulphate, being of different molecular structure, do not even penetrate the skin.

Certain organs of the body have definite protective screens. A brain–blood barrier exists in adults preventing many chemical agents from reaching the brain tissues even though they are present in the bloodstream. In the days of leaded petrol and lead water piping lead poisoning was hardly ever found in adults because of this barrier but was often present in children of inner cities because the formation of this barrier was not complete.

THE LIVER AND THE KIDNEYS

These two target organs perform the vital function of cleansing from the body most of the toxic substances that get into it but can be severely damaged in the process. The liver, for example, is thought to take the brunt of the load of blood purification, and heavy intakes of alcohol, and other poisons, may lead to cirrhosis as this vital organ becomes overtaxed.

Where an acute exposure to toxic substances occurs the extra load on the liver and kidneys arises in a different manner. The first effect of acute poisoning is usually severe stress, similar to shock. The body's circulation and respiration are visibly affected. However, the body systems maintain the blood supply to the brain but reduce the supply to the liver and kidneys, cutting this off in a short time after the exposure. While this protects these organs from chemical damage from the exposure (via the blood circulation) lack of blood to them for any length of time can cause them to atrophy or become otherwise damaged.

This means that in acute chemical poisoning cases resulting from excessive overexposure, treating the person for shock is liable to save their life even though the nature of the poisoning involved may not be known.

'Do not induce vomiting...'

The route of entry of a toxic substance into the body often changes the nature of the toxic effects so the treatment for such a case of exposure is not standard. Trichloroethylene, a common degreasing agent, for example, acts as a systemic poison by ingestion into the stomach but the inhaled vapours mainly cause anaesthesia. So with two modes of entry there are two different effects. This is why the legend 'If ingested, do not induce vomiting' is found on many containers of solvents. Vomiting can allow solvent from the stomach to enter the lungs via the windpipe and the victim may then be subjected to both the above mentioned effects simultaneously.

Instructions on what to do if a person gets any toxic substance onto or into their body *must* be strictly adhered to and should be posted close by the work area where such substances are used.

CARBON MONOXIDE (CO)

Although there are thousands of toxic substances in industry, in society and in the home (many from DIY sources) some are notorious for their killing record. Carbon monoxide is one such killer.

How it kills

The body's haemoglobin (the red oxygen carrying pigment of the blood's red corpuscles) has a strong attraction for carbon monoxide, which is about 200 times greater than its attraction for oxygen. A carbon monoxide concentration in air of 1000 ppm which equates to 0.1% competes for haemoglobin on inhalation on an equal basis with oxygen at its normal level in air of 20.4%. Combining with the haemoglobin, then, the carbon monoxide renders the latter incapable of carrying oxygen to the body tissues, including the brain, and produces an effect of asphyxia.

CO is colourless, odourless and tasteless and in cases of acute exposures collapse of the victim occurs without warning. In chronic exposure cases over a period of years – which do no more than cause headaches at the time – constant starving of the brain of oxygen causes scar tissue which leads to permanent brain damage.

HYDROGEN SULPHIDE (H$_2$S)

This gas, well known for its 'rotten eggs' smell, is one of the most dangerous materials found in industry.

How it kills

Its toxic action is via a process known as 'enzyme inhibition' and this causes respiratory paralysis. One inhalation of a strong concentration can be fatal.

The prime danger of H$_2$S is the fact that although its odour is strong and disagreeable at safe levels, high concentrations cause olfactory fatigue which means you cannot smell it at all. The rule of thumb in an area where this gas may be present is: 'If you smelt it before but you cannot smell it now, get out while you can!'.

GAS ODOURS

In many cases workers use their sense of smell to warn themselves of the presence of dangerous gases or fumes. There is a fundamental rule to bear in mind here.

If you use your sense of smell to warn yourself of danger you should ascertain whether the threshold of smell of the substance is higher or lower than the threshold of danger. If the threshold of smell is *lower* than the threshold of danger, i.e. the concentration which is dangerous, then the existence of the smell is a timely warning that danger is developing. If the threshold of smell is *higher* than the concentration which is dangerous then you should exit that environment as quickly as possible.

ROUTES OF ENTRY OF TOXIC SUBSTANCES INTO THE BODY

There are three routes by which a toxic agent can enter the body: ingestion, inhalation and absorption.

Ingestion

This is the swallowing of industrial products in work situations where employees eat, drink or smoke without washing their hands or where foods or cigarettes are stored near where such products are being used. This is one reason why the law requires rooms where meal breaks are taken to be separate from the actual work areas. Many employees are still their own worst enemy by failing to observe even the most basic aspects of personal hygiene. There is a widespread myth that the body expels unwanted substances by natural means. While this is partly true there are many toxic substances which *remain* in the body slowly damaging vital organs or causing cancers. Death from poisoning is invariably the end result.

Absorption through the skin

This is a more insidious or hazardous route of entry into the body than ingestion because it is not easily noticed and there is so much handling of products, chemicals and raw materials in industry which give rise to it. Dermatitis from skin contact with such materials is the most common industrial disease. Oil and solvent dermatitis cases are more common than acid or caustic burns because employees have a healthy respect for the type of acute injury caused by strong chemicals. However, chronic exposure to oils and solvents, which defat the skin and make it prone to drying, cracking and subsequent infection, does not command the same respect. This is a good example of the difference between toxic substances and hazardous substances in the workplace.

THE BODY'S DEFENCES

The body's defences against the effects of toxic chemicals are very good. They fall into two categories: primary defences, and secondary and third line defences.

Primary defences

- **Eyes** – tear production and blinking which form the natural cleansing process anyway
- **Nose and nasal passages** – sneezing and natural mucus flow. But consider the effects of sensitisers, pollen, hay fever and so on
- **Trachea and bronchii** – ciliary movements and mucus production
- **Throat and lungs** – coughing and throat clearing – expectoration
- **Stomach and intestine** – vomiting and diarrhoea.

Secondary and third line defences

- **Lungs** – expiration of volatile substances
- **Liver** – detoxification by metabolism* and excretion in bile and faecal evacuation
- **Kidneys** – excretion in urine after removal from the bloodstream
- **Skin** – excretion via sweat/perspiration.

*Metabolism: This is defined as the 'process of life' by which tissue cells are destroyed by combustion (katabolism), or the using up of energy by a variety of activities and the renewal of that tissue from chemical substances carried around the body in the bloodstream and derived from digested food (anabolism).

The body's natural defences against toxic attack are very good and the fact that people seem not to come to any harm after exposure on certain occasions tends to instil a sense of false security. However, there are limits to what the defences will stand before collapsing in retreat. Cirrhosis of the liver due to alcohol poisoning is a typical example.

SUMMARY

The material in this chapter is intended to give a broad insight into the threats to employee health posed by exposure to hazardous substances stored, used, produced or handled in the workplace primarily, but not exclusively, for the benefit of the SME manager. So many misconceptions about such substances exist that managers of larger undertakings might clear up many misunderstandings about chemical hazards in their work environments, too. The terminology, as far as possible, has been kept simple, understandable and free from the industrial chemist's jargon.

The mention of the body's natural defences against toxic substances might suggest that prevention of employee exposure to such substances might be relaxed if there is a problem of cost of such prevention. This is a dangerous concept which in these litigation-happy times could cost an undertaking dearly.

Case study 4.2

Compensation paid to 365,000 people claiming they had been exposed to asbestos at work while employed by an American company, Federal Mogul, amounted to nearly US $1 billion since 1982. This huge company, which bought the UK giant, Turner and Newall, in 1998, is only one of 30 companies made bankrupt by compensation claims in America where the compensation culture has gone mad. The latest is that the courts there have so far awarded nine Federal Mogul employees US $1.5 million between them on the possibility that one day they could get sick from an asbestos related disease. So far none of the nine show any symptoms whatsoever of such illnesses.

Many cases of employee exposure to toxic substances result from ignorance, carelessness, indifference or downright laziness – 'They can't be bothered to use the personal protective equipment' is a frequent reason for an exposure. Also, because there is often no immediate *reaction to the exposure those involved think that no harm has been done when in fact an insidious build-up in their bodies of the substance(s) concerned is occurring.*

The Control of Substances Hazardous to Health Regulations have, for some years now, set out the basic legal requirements to control exposure of employees to toxic and hazardous substances at work. These regulations are relatively straightforward and represent a good deal of common sense as far as preventing occupational ill health is concerned. The main thing for the SME manager is to know what dangerous properties a substance possesses and what harm it can do to the human body in the course of its use, storage, handling, transporting and processing. This information is free from the suppliers (who are bound to provide it by law) and from the HSE Infoline 0870 154 5500.

CHAPTER 5

Surveying, monitoring *and* sampling techniques

It has been seen that a wide range of potential health hazards can be posed by hazardous substances in a workplace.

It is, therefore, necessary for the SME manager to be aware of the surveying, monitoring and sampling techniques at his or her disposal to enable the presence of, for example, airborne contaminants to be detected before they reach dangerous concentrations.

WHERE TO START

A starting point for this activity which requires no specialist knowledge is a survey of the medical history of all the employees in the areas where hazardous substances are present, for whatever reason. Is there any trend or pattern of excessive absenteeism or regular illnesses? If there is, the reason(s) for this should be determined. Mention has been made already of the legal requirements under COSHH to control exposure of employees to substances hazardous to their health. However, the regulations do not describe the effects of exposure on the human body to any great extent; that is left to the vast amount of literature – much of it free – published by the HSE often in leaflet form.

It is advisable for employers to be aware of the effects on the body of hazardous substances present in their work environments and not to simply confine their knowledge to whether a substance is toxic, very toxic, irritant, harmful, carcinogenic and so on. As mentioned above, the HSE produce information on every hazardous substance an undertaking, and certainly an SME, is likely to have on its premises. In addition, suppliers *must* provide safety data on substances used at work. The full text of this legal requirement is contained in section 6(4)(c) of the Health and Safety at Work Act and this is the primary source of information for all managers. The main benefit from this source is that the manager gets only the essential information and is not swamped with superfluous data about substances not present on the premises. Many firms purchase very detailed and often expensive sources of chemical hazard information only to find that they require only a small fraction of the information they have purchased.

INFORMATION A SUPPLIER MUST PROVIDE

There are still a number of suppliers who fight shy of providing full information about their products. Most claim 'commercial confidentiality' in the first instance but what they really mean is that their products contain substances that might cause concern to clients who discover these for the first time.

The supplier must provide *specific* information on their products. It is not sufficient simply to draw attention to the dangerous nature of the substance. The supplier must ensure that the user is aware of the risks to which the substance may give rise, the circumstances in which the risks may be accentuated and the means whereby they may be reduced to a level which does not jeopardise the health and safety of the user's employees. The HSE has also stated that the information must be 'adequate', and that this means that the client must be able to understand it. If he or she can't then the information is not adequate.

```
Case study 5.1

A large College of Art and Design was in receipt of
an Improvement Notice from the HSE because of
inadequate COSHH assessments for the vast range of
substances hazardous to health held in its
laboratories, studios, classrooms and so on. Whilst
the quantities were invariably small, the aggregate
number was quite formidable. Information from
suppliers was patchy and often inadequate.

The College set about obtaining fuller information
from suppliers but one supplier of photographic
materials was adamant in refusing to identify the
ingredients of a product which was simply labelled
'Dangerous. Do not allow to come into contact with
skin. Do not ingest'. He claimed that the formula
was subject to commercial confidentiality. Only
when he was threatened with prosecution by the
Inspector did he reveal the contents of the product.
The irony of the situation was that the product
comprised common ingredients which could be obtained
from most High Street chemists. It was being sold
under a trade name at an inflated price and was
classified under COSHH as 'irritant'.
```

TRADE NAMES

Trade names are used for a variety of reasons. They can form a source of advertising for the manufacturer, e.g. 'Shellsol' and 'Dowpon', or they can refer directly to the job the product is intended to do. Also, a product may contain a variety of different chemicals so a single name is convenient and easy to remember. Unfortunately, however, many chemical products are sold under trade names to keep the compositions confidential, although this is breaking the law. In such cases it is difficult to find out exactly what chemicals are used in such products.

An employer needs to have full information on trade name products for the following reasons:

• to choose the least toxic product capable of doing the job
• to provide adequate control measures
• to know what first aid measures to provide in cases of overexposure
• to be able to carry out a cross check where necessary on the accuracy of the suppliers' information.

It is also essential that the employees are aware of the true toxic properties of all trade name products used in their work environment.

DANGERS OF SUBSTITUTING 'SAFER' SUBSTANCES

It is generally thought that the substitution of a safer, less toxic substance for a more toxic one is good practice. Generally this is the case. However, replacing a product with a substitute may reduce, but not eliminate, the hazard.

THE IMPORTANCE OF GOOD HOUSEKEEPING

A messy, dirty and untidy work environment is a unhealthy work environment even before toxic substances are introduced into the equation. Where chemicals *are* present in such situations it is unlikely that spillages are cleaned up and the chance of dermatitis due to the vapours of solvent spills is but one example of ill health hazards. This is apart from increasing other safety problems as well.

The physical layout of work areas is another important aspect. Certain operations should not be carried out too close to each other. For example, if electric arc welding or hot cutting or burning is being carried out this must not be close by an area where halogenated vapours from degreasing operations are present. The intense heat from the hot work can pyrolyse chlorinated hydrocarbons to form, among other things, highly toxic phosgene gas (suppliers' safety information should state clearly any hazards created by a product's exposure to heat).

OCCUPATIONAL EXPOSURE STANDARDS

These standards were originally called 'Threshold Limit Values' (TLVs) and then 'Occupational Exposure Limits' (OELs). They are now called 'Occupational Exposure Standards'. All mean the same thing and define the concentrations of hazardous substances in air, averaged over a specific period of time – normally 8 hours – below which the average healthy worker will suffer no demonstrable ill effects. However, there may be cases where, because of age, constitution, climate or variation in human susceptibility, exposure to a toxic substance at, or even below, the OES may cause discomfort or illness. Occupational Exposure Standards should, therefore, be regarded *only as guide values* in the control of hazards and *not* as fine dividing lines between safe and dangerous concentrations.

Case study 5.2

(This case study is taken from the US National Institute for Occupational Safety and Health, *Working in Confined Spaces, Criteria for a Recommended Standard*, 1979, annals. It could just have easily occurred in the UK).

Trichloroethane was substituted for the more toxic trichloroethylene at a radiator and metal tank repairing and cleaning firm. The worker would saturate a pad with the solvent and lower himself head first into a tank that needed cleaning and clean the inside of the tank as quickly as possible. A worker following this system of work was found with his legs protruding from the upper manhole of the 450 gallon tank and was unresponsive. He was removed immediately and given artificial respiration until a doctor arrived and pronounced him dead. A reconstruction of the fatal accident revealed that the concentration of trichloroethane vapour in the tank had reached 62,000 parts per million (ppm). The exposure concentration permitted by HSE's EH40 is 200 parts per million. It transpired that the workers had assumed that since the substitute cleaning solvent was less toxic than the one previously used there was less danger. The exposure concentration of trichloroethylene is 100 ppm. The deceased worker was exposed to 320 times the occupational exposure limit for trichloroethane.

Once it is established that the presence of such substances is unavoidable then the personal exposure levels must be determined to see if they are acceptable. Wherever possible, these levels are determined at the breathing zone of an employee for an inhalation hazard, rather than simply at the workstation (a similar requirement applies to noise measurements: these should be made as near to the employee's ear as possible and not simply at the workstation).

For airborne chemical hazards some sort of air sampling and analysis is required. The quickest, or 'spot check' method uses detector tubes. These are transparent tubes about 5" (12 cm) long and $\frac{1}{4}$" (5mm) diameter containing chemicals which change colour or extend a stain in proportion to the amount of a specified compound drawn through the tube by means of a hand held pump. The stain tubes may be specific for one compound (NO_2, CO, NH_3 etc.,) or for a category of compounds (alkenes or olefins). Their accuracy is a subject of continuous improvement by manufacturers, and can vary by up to 25% under certain circumstances.

In practice during the working day the concentration of an airborne contaminant can vary and it normally fluctuates around a mean value. The amount by which the OES can be exceeded for short periods without injury to health depends on various factors such as the nature of the contaminant, whether or not it has cumulative effects, whether exposure to very high concentrations, even for short periods, can cause acute poisoning, the frequency with which the OES is exceeded and the duration of the excessive concentration.

IS THE ENVIRONMENT DANGEROUS?

All the above factors need to be taken into account in deciding whether a situation is dangerous or not. In general, industrial hygiene practice has tended to keep exposure levels *below,* rather than at, the OES. Overall, detector tubes must be recognised as screening devices rather than sophisticated sampling/analytical ones.

Similarly, Occupational Exposure Standards have been established on the basis of the best information available and although they are presented in terms of numerical standards, they do not represent a precise definition of a hazard/non-hazard situation. They are subject to annual revision and do not indicate the values below which all individuals will be protected and above which harm will befall everybody. There will be cases where some employees will suffer adverse effects at exposure levels well *below* those set by the OES. These standards must, therefore, be treated as good practice guidelines only and have been accepted by the HSE as such.

Another common misunderstanding amongst users of stain tube detection techniques is the lack of understanding that OESs are based on long term *average* concentration values. When instantaneous grab sampling techniques are used to determine concentrations in work environments it is to be expected that there will be short concentration periods (known as 'excursions') above the OES.

HSE PUBLICATION EH40: OCCUPATIONAL EXPOSURE LIMITS

EH40 is one of HSE's Guidance Note series which basically provide Occupational Exposure Standards and Maximum Exposure Limits for hazardous substances in tabular form. Both types of limit are concentrations of hazardous substances in air averaged over a specified period of time referred to as the Time Weighted Average (TWA). Two time periods are used: long term (8 hours) and short term (15 minutes). The Short Term Exposure Limits (STELs) are set to help prevent effects which may occur following short exposures of only a few minutes.

Maximum Exposure Limits (MELs) are set for substances which may cause very serious health effects such as cancer or occupational asthma and for which 'safe' levels of exposure cannot realistically be determined. They also apply to substances for which safe levels may exist but where control to those levels is not reasonably practicable.

Occupational Exposure Limit values are stated in parts per million (ppm) or milligrams per cubic metre (mg.m^{-3}).

The use of detector tubes for sampling air quality in work environments has both benefits and drawbacks. The benefits are that the stain tubes are direct reading, relatively inexpensive and can be used by comparatively unskilled persons after a few minutes' training. The drawbacks include the fact that people who *do* use them frequently do not fully understand the principle of the Short Term Exposure Limit

and the fact (already mentioned above) that tubes are, at best, only accurate to ± 25% .

Nevertheless, just as the cheap, pocket sized version of the industrial sound level meter gives an early warning of existing or developing noise problems in a workplace, the stain tube detection method gives a good, clear indication of the presence or otherwise of hazardous airborne contaminants. The tubes are manufactured to cover different scales so that trace elements of, say, ozone can be detected which might not show up on a tube designed to measure larger concentrations. Stain tubes are invariably used to test for gas in confined spaces prior to entry. It has to be stressed once again, however, that each tube is normally for a specific substance so one has to know what substance one is testing for.

Examples of OELs, MELs and STELs
OEL

Ammonia NH$_3$	25 ppm	16 mg.m^{-3}	STEL 35 ppm	25 mg.m^{-3}

MEL

Dichloromethane CH$_2$Cl$_2$	100 ppm	350 mg.m^{-3}	STEL 300 ppm	1060 mg.m^{-3}

Note: where fibrous airborne contaminants are concerned, e.g. asbestos, concentrations are measured in fibres per millilitre of air.

This chapter has stressed the need for employers to gain as much information of toxic substances used on their premises as possible and has shown some of the methods by which such information can be obtained. It has also explained how hazardous airborne gases, aerosols, vapours and so on can be detected and how their concentrations can be measured. Amongst other things it is imperative that full information about a toxic substance an employee may have ingested or inhaled is available when medical help is summoned so the correct treatment can be administered right away. Delay through incomplete knowledge can be fatal. Do not expect a doctor to have the information.

Hazardous dusts

Occupational ill health due to harmful dusts is usually associated with the respiratory tract and the lungs. However there are dusts composed of poisons which are absorbed into the bloodstream and cause damage to other parts of the body.

The word 'dust' has no precise scientific meaning; it is usually taken to mean a solid which has been broken down into a 'powder', so that the individual particles are too small to be distinguished easily. Both the danger of dust to health and the way it behaves in air are dependent very much on the size of the particles (this point is very much a deciding factor in dust explosions, too).

The Health and Safety Executive requires any dust concentration in a workplace to be below 10mg.m^{-3} regardless of particle size and the constituents of the dust.

PARTICLE SIZES AND RESPIRABLE DUSTS

Dust particles are measured in 'microns', these being one thousandth of a millimetre. The smallest particle visible to the naked eye is some 50–100 microns in diameter. In the main it is the very small particles – far too small to be seen – which are the dangerous ones, but with some dusts e.g. asbestos, cotton and agricultural dusts, the larger particles can be dangerous, too.

The apparent size of a gas molecule in air is 0.003 microns. The most dangerous dust particle size for the lungs, known as 'respirable dust size', is normally between 0.2 and 5 microns. Particles larger than 5 microns tend not to penetrate the lungs and those smaller than 0.2 microns settle so slowly they seldom reach the lung lining. Since the dangerous dust particle size is invisible to the naked eye, *visible* dust gives only a warning of danger, as where there is visible dust there will certainly be invisible particles also. However, the fact that no visible dust is present does not mean that the dangerous invisible dust is not present.

DEPOSITION OF DUST IN THE LUNGS

The amount of dust deposited in the lungs depends upon the balance between deposition whilst breathing in and removal on breathing out. The actual deposition in the lung depends on particle size, shape, density and upon the depth and rate of breathing and whether the breathing is through the nose or the mouth. Some 55% of respirable dust (0.2–5 microns) is retained in the lungs and the respiratory passages.

Compared with the area of the nose or mouth in respect of breathing in the area of lung lining which interacts with the circulating blood and the inspired air is very large – some 80 square metres. As inspired air along with dust particles travel down the respiratory tract, the velocity of air movement slows down. Also, in the wider air passages of the upper respiratory tract, the airflow includes a degree of turbulence which, through centrifugal force, throws the larger particles of dust out of the airstream. These are cleared from the throat areas by expectoration. Further down in the lower, narrower passages the air flow is slow and non-turbulent so deposition of dust particles onto the lung linings depends on a number of other factors.

WHAT HAPPENS WHEN RESPIRABLE DUST PARTICLES REACH THE LUNGS?

The construction of the walls of the lungs is highly complex, but in simple terms, the walls are made up of tiny sacs called alveoli. These sacs are responsible for effecting the gaseous interchange where oxygen breathed in is passed through the sac walls into the bloodstream and carbon dioxide is removed from the bloodstream and expired. Dust particles reaching the alveoli affect their proper functioning and thus affect normal breathing. Differing chemical dusts (silica, coal dust, asbestos dust, dusts of heavy metals, beryllium dust and so on) cause different types of lung disease but the one end result is always the same – difficulty in breathing as the area of healthy lung is slowly damaged irreparably (there is a parallel here with noise induced hearing loss caused by the 'hairs' in the cochlea being killed off by excessive noise).

As the number of healthy, working alveolar sacs reduces an additional strain is placed on the heart to pump blood through the lungs. The condition is called *emphysema* and is often associated with chronic bronchitis. It is a common condition accompanying many occupational dust diseases and causes breathlessness on even mild exertion or even at rest.

DISEASES CAUSED BY DUSTS

These are known collectively as 'pneumoconioses' (plural). The most commonly known example, particularly in the old coal mining areas of the UK, is probably coal dust pneumoconiosis. There are two groups of pneumoconioses; benign, with little or no fibrous tissue reaction, and fibrotic, due to what are termed 'proliferative' dusts.

Examples of dusts causing benign pneumoconiosis

1 **Carbon** dusts e.g. smoke, soot, carbon black (but some soots can be carcinogenic). Not visible on X-ray.

2 **Calcium** dusts e.g. cement, marble, gypsum. Not visible on X-ray.

3 **Iron** dusts e.g. from welding, grinding, mechanical cutting of steel. Causes chronic inflammation of the lung, a condition known as 'siderosis'. Opaque shadows visible on X-ray.

4 **Artificial abrasives** e.g. carborundum, aluminium oxide from grinding wheels.

5 **Barium and tin** dusts give opaque shadows visible on X-ray.

Examples of dusts causing fibrotic pneumoconiosis

1 **Silica** e.g. free crystalline silica as in quartz, flint. Produces fibrosis in which collagen* is deposited in nodular clumps throughout the lung. Often complicated by presence of calcium, iron, aluminium and carbon dusts. The disease is known as 'silicosis', can lead

to emphysema, and predisposes to pneumonia and tuberculosis.

2 **Asbestos** produces diffuse fibrosis mainly in the lower parts of the lungs. The dust has the effect of causing a collar-like growth around the end of the smallest bronchial tubes of the respiratory system thus closing the off the supply of air to the alveoli sacs and thus to the lung lining. Eventually, after 4-7 years of exposure, symptoms of emphysema develop and clubbing of the fingers develops. Clinically, the most striking sign is increasing shortness of breath accompanied by a persistent dry cough.

3 **Silicates**. The dust of certain silicates e.g. talc, kaolin and mica can produce fibrotic changes in the lungs.

4 **Bauxite or alumina**. In the fine white powder form (aluminium oxide) can cause lung damage in some cases. The condition is known as Shaver's disease.

5 **Coal dust**. Pathological changes in the lung depend on the type of coal dust. Miners of soft coals usually develop benign pneumoconiosis whereas fibrotic pneumoconioses are associated with the mining of hard coal. This has a higher silica content which is thought to produce the fibrotic condition which leads to emphysema and shortness of breath.

6 **Beryllium compounds**. Beryllium is used in the manufacture of fluorescent lamps and tubes and is a potentially dangerous substance from a respiratory disease point of view. The dust can cause severe inflammation of the lungs and an acute type of pneumoconiosis in severe cases. Mild cases can be cured but severe exposures can be fatal.

* *Collagen: a protein constituent of fibrous tissue; in simple terms, a growth associated with areas of tissue which is fibrotic.*

DUST CONTROL MEASURES

Measures for assessing concentrations in air of gases, vapours, mists etc., by air pump and stain tube methods, are not feasible for measuring concentrations of particulate matter such as dust. This makes dust sampling much more complex and expensive. With dust, measurements have to be taken which determine the concentration by weight per unit volume (gravimetric sampling) or by visible counting of fibres, e.g. in the case of asbestos, under a microscope.

Gravimetric sampling equipment is fairly compact and one version can be attached to a person at work thus enabling their personal exposure to be measured as they go about their work. The device comprises a battery operated pump which draws air at a preset volume flow rate through a preweighed filter on which the dust is deposited. Reweighing the filter after a set time will indicate the amount of dust collected and the concentration can then be calculated.

The best possible way of controlling dust is to stop making it in the first place. This is obviously the most difficult way but where it can be done it offers a final solution to the problem. It may involve changes to systems of work or manufacturing methods; the biggest problem here can be management inertia because dust is wrongly viewed as a mere nuisance and nothing to really worry about.

Case study 6.1

A woodwork teacher in a Welsh secondary school
recently won £200,000 after claiming that sawdust
in the school workshop ruined his health. He claimed
exposure to sawdust in the workshop which was not
cleaned properly gave him a dust allergy, allergic
rhinitis and asthma. Repeated representations to
his Local Authority employers had no effect.

The HSE drew attention to the fact that all wood dusts (including softwood, hardwood and wood composites like chipboard and fibreboard) are hazardous to health in 1997 in its woodworking information sheets nos. 11 and 14. Target organs are the nose (nasal cancer), respiratory system and the skin. Hardwood dusts are particularly harmful and the COSHH Regulations set a maximum exposure limit of $5mg.m^{-3}$ for such dusts.

If a dust cloud cannot be completely eliminated then as much of it should be contained as possible. Where powder creates the dust, if it can be processed in suspension in a liquid, in solution or moist, dust is less likely to be present. Coal dust suppression measures lately included fine water spraying close by the teeth of the face cutting machines to contain the dust in a fine slurry. However, moistening of powders is not always feasible, particularly where machinery is concerned, because of clogging.

The use of synthetic grinding wheels instead of old fashioned sandstone has eliminated free silica particles that result from the breakdown of the wheel in use, but some sandstone wheels are still used in the agricultural industry.

Undertakings such as animal feed provender mills and flour mills can generate considerable amounts of dust if the machinery and plant are not well maintained. These mills elevate raw material, which must be dry and therefore dusty, to the top floor of quite high buildings into different hoppers. From here the materials are fed through agitators, mixers, cyclones and other plant from where eddies of dust emerge freely from orifices in worn plant. In such premises, dust can gather in quite large quantities if regular cleaning and maintenance of plant are not carried out. Apart from health considerations the possibilities of dust explosions (already mentioned) are always present.

TYPES OF VENTILATION

Much dust, rather like welding fumes, can be extracted at source by vacuum systems and Local Exhaust Ventilation (LEV). Such systems can incorporate material recovery features for recycling purposes. Local Exhaust Ventilation installations have to be properly designed and maintained, otherwise they are a waste of money and time. The essential parts of such a system to be considered are:

- the source of the dust
- the hood, i.e. the mouth of the system through which the dust is extracted
- the ducting itself
- the air cleaning plant
- the extractor fan.

The main thing to remember about such extraction methods is that the extractive force being drawn into an opening decreases significantly with the distance from the open face. This loss of extract velocity thus requires the hood to be positioned as close as possible to the source of the dust.

This loss of 'capture' velocity is amply demonstrated by the comparative uselessness of 'extractor' fan units mounted in the ceilings of bathrooms and toilets and in kitchens and pub lounges. These rarely perform well because they only ventilate a very small local area.

Another type of ventilation which still extracts airborne pollutants is *plenum* ventilation. In this system clean air is 'pushed' into the work area at point remote from where the main extraction plant operates. Normally any form of extraction of air from other than a sealed area will result in replacement air being drawn in to replace that being extracted. Plenum ventilation assists this process and at the same time has the effect of diluting the concentration of dust in the air anyway. Where plenum ventilation is employed care must be taken to ensure that the velocity of the incoming airstream does not deflect contaminated air from the capture area of the LEV hood.

Should LEV be found to be necessary, have the installation designed and installed by professionals in this field. Ventilation is a skilled job and much time, money and effort can be and is wasted because the work is given to people who are not sufficiently skilled to understand the fundamental physical and engineering principles involved.

Where processes create dust which cannot be effectively controlled by any of the above means and it passes into the general atmosphere of the room or work environment, then all that can reasonably be done is to flush the area with clean air as often as possible. This method is a last resort, as it will be seen that in winter and during inclement weather the necessary measures to effect this 'flushing', i.e. opening of doors, windows and so on, will not be popular with employees working there. Flushing in this manner will remove heat, and bringing the work area up to a comfortable and legally required temperature can be expensive. In a large metal fabrication shop which had problems of noise, lighting and welding fumes – the shop was fully open plan – the management said its prime dilemma was due to the fact that if the roof extractor fans were switched on the air was cleaner but the temperature was lowered considerably. If the fans were off the atmosphere was comfortably warm but polluted to a very high degree. This problem arose only during the winter months.

Dust is a greatly underrated hazard at work which many employers fail to take seriously. Like many examples of exposure to substances hazardous to health the adverse effects of over-exposure are usually chronic as opposed to acute. It is still the most widespread of the industrial killers and has been known to be so for many years. Chief Inspectors of Factories Reports published long before the 1974 Health and Safety at Work Act stressed the importance of dust control measures and a Report of 30 years ago drew attention to the apparently intractable and ever present problems of industrial dust which have defied control for more than 100 years.

However, as has been shown above, dust *can* be suppressed and controlled in many workplaces but it is often expensive, time consuming and requires expert advice to do so. Also, some dusts are deadly as far as worker health is concerned; others are relatively harmless at controlled exposure levels. Most dust concentrations are anything but pure and form a veritable cocktail of different types. This introduces the question of dust sample analysis, the costs of which, to the SME manager, can appear astronomical in relation to the work actually carried out. However it is vital to identify the main *types* of dust present in a work environment to reduce or eliminate the hazard. The ILO identifies four categories of dust on the basis of their effects on the body. These have been identified above but are summarised below for convenience:

- **Inert dusts** – accumulate in the body producing no ill effects. In the lungs may impair lung clearance by obstructing lymph* flow.

- **Toxic dusts** – often metal compounds such as chromates and lead but can be mineral e.g. asbestos. Can have acute or, more often, chronic effects on target organs.
- **Allergic dusts** – can cause eczema or asthma. Sometimes difficult to define from an occupational health point of view.
- **Fibrogenic dusts** – most dangerous. Lead to the pulmonary diseases, pneumoconioses and emphysema. Difficulty in breathing and shortness of breath results.

lymph: the fluid from the blood which has passed through capillary walls to supply nutrients to tissue cells.

GOOD HOUSEKEEPING

Once again the best, and cheapest, measure to achieve and maintain a healthy and, in this case, a dust free work environment is good housekeeping. In case study 6.1 above regarding the woodwork teacher winning £200,000 compensation for sawdust problems, it is difficult to comprehend the reasons why a workshop could not be kept clean when sawdust was the primary contaminant. Whilst sawdust doubtless contributed to the teacher's clinical conditions, the root causes of the dust problem probably included lack of will, ignorance, laziness or job overload on the part of those charged with the cleaning of the workshop. Sawdust is not exactly difficult to clean up when one considers the amounts generated on a school workshop!

All accumulations of dust should be removed – in fact, so far as is reasonably practical, the accumulations should be prevented in the first place. The layouts of furniture, plant and so on in areas where dusty processes are likely to take place should be such as to enable cleaning operations to be carried out thoroughly and the floors of such premises should be free from obstructions.

It is obviously unrealistic to keep a production area, or any work environment, as clean as an operating theatre should be kept, but the less dust that is present, of any kind and from any source, the better. Not only will employee health be better but there are fewer chances of claims for compensation for dust related diseases in the future.

SUMMARY

Examples of dusts from the above categories have been provided in this chapter and their effects on the body described. 'Safe' concentrations have been described and methods of suppression and elimination dealt with. Since the effects of dust on health are usually cumulative, it is important, so far as is reasonably practicable, to minimise the length of time during which any one person is exposed to dust, even at low concentrations. This is seldom done. Ideal though it may be, it is practically impossible for workers to be employed in dust free areas for most or all of their working day. Only too often, workers are continuously employed in such places where high levels of dust exist, for example, operating controls, weighing batches, bagging off in materials handling plant or manually mixing powdered compounds, often in the very worst locations.

With present-day automation this exposure can be avoided but often at a cost beyond the reach of the SME. But some measures can be quite inexpensive; sometimes controls can be taken to a computer controlled console in a properly ventilated cabin. Staff still have to enter the dusty areas from time to time but if periods of entry are relatively short, risks will be proportionately reduced. Chapter 3 on 'Noise' described how removing hearing protection for even a few minutes in a noisy work environment nullified the protection for the rest of the day, even if the protection was used for that remaining time. Dust damage is generally cumulative and a person exposed for, say, one hour will only get half the effects as, say, a person exposed for two hours.

CHAPTER 7

Occupational stress

With the current change in emphasis away from the traditional manufacturing industries towards service industries, often based on information technology, occupational stress is an area on which safety practitioners and advisers are increasingly having to turn their attention.

INTRODUCTION

Once regarded as something of a joke in the world of work, stress is now treated seriously; it is a major subject of contemporary compensation culture and is the topic about which volumes of rhetorical nonsense are written. One recently published article about 'managing stress' in an organisation suggests making middle managers 'the intended foci of action' for a stress management programme, conveniently overlooking the fact that it is often these very managers who are the main sufferers from stress in the organisation itself.

Stress is people's natural reaction to excessive pressure placed on them when they are not in control – rather like involuntary risk where the employer, not the employee, dictates the risks the latter has to take. Stress is not a disease but if continued for any undue length of time it can lead to physical and mental ill health. Depression, nervous breakdown, suicidal tendencies and heart disease can be manifestations of stress. People often dismiss stress as 'being good for you'. Being under a degree of *pressure* often improves alertness, initiative and performance generally. But when pressure becomes excessive it leads to stress and stress is actually *bad* for people. For the employer there are the true costs of stress, just like there are true costs of accidents. Frequently the employer is unaware of each of these.

SOME FACTORS REGARDING STRESS

Everybody is affected by stress but people react differently to it. Different people obviously have different stress thresholds and tolerance levels to stress. As mentioned above, a certain amount of stress is good for us as it improves efficiency, But beyond a certain point it becomes destructive.

Stress is the body's automatic response to real or imagined danger or challenge. This challenge can arise quickly or gradually over a long period. Physiologically, when faced with a sudden threat such as a redundancy announcement late on a Friday afternoon, the reaction is to fight the threat or run away from it. Symptoms of stress include increased heartbeat, increased breathing rate and dilation of the pupils of the eyes. Adrenaline flow speeds up to allow the person to cope with the sudden stress. This leads to faster blood circulation providing more oxygen to our muscles. The blood sugar level rises, providing larger amounts of disposable energy.

WORK RELATED STRESS

Work related stress is stress which is caused, or made worse, by work. It is a feeling of distress associated with extreme levels of physiological arousal and often feelings of helplessness. Some of the factors which contribute to work related stress include:

- **Boring or repetitive work or too little to do.** Can job rotation be introduced? Can the system of work be varied in any way to reduce the boredom? Can more work be provided?

- **Too much to do in too little time.** A common problem. Can a redistribution of work be effected? Can the way jobs are done be changed? Can more warning be given of urgent jobs? Can tasks be better prioritised?

- **Confusion about exactly how everyone fits in.** Everyone must have clearly defined duties and responsibilities. They should know the duties and responsibilities of their immediate seniors and subordinates, too.

- **Having responsibility for others**. In far too many situations workers are told to 'inform their supervisor or manager' when an unsatisfactory situation develops. All too often that supervisor or manager is not trained to deal with the matter or, invariably, is far too busy anyway. Where these instructions exist, the supervisors and managers concerned must (a) know what to do, and (b) have the time and resources to do what is required.

- **Poor relationships with other work colleagues.** A rather glib answer is frequently given in these situations such as 'provide training in interpersonal skills'. This sort of training is highly specialised, expensive, often lengthy and not always guaranteed to work (the author has had experience of this). A frank, sincere and completely confidential discussion with the person concerned, who often feels alienated, is a more pragmatic approach.

- **Bullying, racial or sexual harassment.** Bullying can take a variety of forms and can be physical or mental. Sarcastic comments about an employee's abilities by a senior person in front of others is a common form. Bullying in any form should not be tolerated but those exposed to it should realise that the best way to stop it is to stand up to the bully. This also applies to employees who feel they are being racially or sexually harassed. The reality is that such harassment is human nature for those who indulge in it and no amount of legislation will change the basic attitude of such people.

- **Inflexible work schedules.** Many sources of rhetorical advice suggest seeing if there is scope for making work schedules more flexible but do not suggest what to do if this cannot be done. There are many types of work schedules which simply cannot be made more flexible and the blunt truth of the matter is that these situations must be coped with.

- **Physical hazards of the workplace.** These can normally be controlled and, in fact, the law requires this – but only so far as is reasonably practicable.

- **Lack of control over work activities.** It has always appeared logical that employees work better if consulted and their ideas on planning and organising their own jobs are at least considered. People feel flattered and pleased when asked for their opinion or for advice.
- **Lack of communication and consultation.** Consultation has been dealt with above but poor communication seems endemic in many organisations. There is no valid excuse for this and improving communication poses no insurmountable problems.
- **Negative cultures and cultures of blame.** These days there is a widespread reluctance to accept blame or responsibility when things go wrong. Everyone should be *accountable* for the responsibilities and duties which go with their job. Amongst other things, this is what they are paid for. Senior staff should be honest, set good examples and listen to and respect subordinates.

SOME COMMON SYMPTOMS OF STRESS

You feel that you are...

not really understood

unable to finish your tasks

neglected; unwanted, undervalued

tense; unable to show your feelings

always angry or irritable

lacking in confidence

unable to cope

in constant fear about the future

...and you might

bite your nails

start or increase smoking

start or increase drinking

snap at people

start compulsive eating

break down; start crying

lash out at someone (verbally or physically)

start taking tranquillisers or drugs.

The above symptoms can result in your suffering from:

- high blood pressure
- constant fidgeting; nervous twitching
- frigidity/impotence
- excessive sweating
- food cravings
- lack of appetite
- constant tiredness
- insomnia
- constipation or diarrhoea
- frequent indigestion.

If any of the above conditions are abnormal in the context of a person's daily lifestyle they are undoubtedly suffering from stress. The normal procedure would be to visit a GP, but GPs normally treat only the symptoms as they cannot be expected to treat the root causes of occupational stress. The fact is, though, that left untreated, stress can cause coronary heart disease, diabetes and bronchial asthma. When a person is under stress, the body is not at ease with itself; the symptoms prevent relaxation which is essential for good health. Stress symptoms can be either physical or psychological. The latter cause organic reactions within the body resulting genuine physical and mental illnesses.

WHAT CAN BE DONE?

The pace of life in general is increasing at such a rate that there is less time to genuinely relax and avoid the common factors that cause stress.

The position at work is complicated by the fact that much stress is domestic in origin; the job simply exacerbates the problem. While managers are not under any legal duty to prevent ill health caused by stress due to problems outside work, these problems can make it difficult for people to cope with work pressures with which they would normally have no problem. Being understanding to staff in this position would be in your interests.

Where stress caused or exacerbated by work could lead to ill health the manager must carry out a risk assessment of the situation. Such an assessment would follow the normal procedure, i.e:

- identify the problems which are capable of causing the stress. If these are not immediately obvious *ask* people what they think
- decide who is most likely to be affected or harmed by these problems
- implement measures to control the problems.

This sounds simple enough and it is. The manager may have to encourage the employee to confide in him or her because symptoms of stress do tend to be regarded as signs of weakness but many of the outward signs of stress are quite noticeable. Changes in a person's behaviour and general demeanour can be spotted. Irritability, indecisiveness, absenteeism and reduced performance are all indicators of stress. Persons who smoke and drink will invariably increase their intake and may even turn to drugs.

As regards the controls mentioned above there is no single best way of tackling the problem. Solutions will depend to a large extent on existing systems of work and the actual causes of the stress. These must be clearly identified, just as risks posed by workplace hazards are identified, and dealt with. Simply talking about the problem, providing advice and sympathy is merely a palliative. The manager must *empathise* with the sufferer and get to the root of the problem.

AVOIDING COMPENSATION CLAIMS

Probably the first significant stress claim to succeed was the one involving a Mr John Walker and Northumberland County Council. In this case Mr Walker had his workload progressively increased and his employer appeared to turn a deaf ear to his representations about being increasingly unable to cope. Several subsequent cases have been successful where the employer has simply failed to *listen* to the complainant or has listened but has failed to act.

While some cases of stress are due to the fact that the employee is simply not up to the job these are in the minority. The fact is that it is better to identify and deal with alleged stress cases long before the employee makes a claim.

As is so often the case when the 'experts' get on board, advice on 'managing stress' by devising 'corporate stress policies' with nominated staff as 'focal points of stress strategies' and similar jargon would, if only half implemented, leave the undertaking practically no time to achieve the objectives it has set for itself. Stress management should be kept simple; for the SME, the management should show that it is serious about the subject. Management can usually keep an eye on its staff and relationships should be such that employees are not discouraged to speak up about their personal difficulties. Managers can

45

and should lead by example and should remember that if one of the workforce is suffering from stress this may be the tip of an iceberg.

SUMMARY

This chapter on occupational stress has tried to explain what stress at work is; how it can be caused; how it can be recognised and what can be done about it. The HSE states that slips, trips and falls form the majority of lost time accidents at work. It will also confirm that a wholly disproportionate amount of time off work for stress and stress related illnesses is presently the norm.

Managers often ask whether they are obliged to provide stress management training for employees. There is no legal obligation to provide this. In any case, each stress case is an individual one and stress training does not necessarily tackle the **causes** *of stress at work. Stress management training might be appropriate for a large undertaking but it is not generally considered worthwhile or economical for an SME. Nor should counselling be entertained under normal circumstances. Counselling has grown into quite an industry in recent years. The author's County Council has even offered counselling to a lady who had a bottle of milk stolen from her doorstep. Employees should be encouraged to discuss their problems with their employer, who is not obliged to provide counselling services (which the employer would have to pay for) and who is best placed to deal with the problem.*

Consultants should also be avoided. They are primarily interested in offering complex and expensive risk management and stress reduction strategies which simply are not necessary to tackle the problem. If any further assistance is required the HSE publish an excellent free booklet **Help on work-related stress. A short guide INDG281** *which certainly addresses all the likely requirements of an SME and much larger organisations as well.*

Legal aspects *of* occupational health *and* hygiene

Prior to the Health and Safety at Work Act in 1974 occupational health matters were regulated by a motley collection of industry, process and substances-specific regulations, orders, Acts of Parliament and so on.

INTRODUCTION

Since 1974, however, the situation has been rationalised by two principal pieces of legislation, the Control of Substances Hazardous to Health Regulations 1994 (known colloquially as the COSHH Regulations) and the Chemicals (Hazard Information and Packaging for Supply) Regulations 1994 (known as the CHIP regulations). These regulations between them provide for the protection of people, mainly at work, against any substance, natural or artificial, which is capable of causing adverse health effects or disease arising from work activities. The substances covered can be in a solid, liquid, gaseous or vapour form and micro-organisms are included.

The COSHH Regulations really should be called 'The Control of Exposure of People to Substances Hazardous to Health' because that is the regulations' intention. However, this would make their title rather unwieldy so the existing title prevailed.

THE CONTROL OF SUBSTANCES HAZARDOUS TO HEALTH REGULATIONS 1999

The 1999 version of the Regulation incorporates all amendments to date – these have all been relatively minor since the Regulations were first promulgated. These comprised the most significant new legislation to emerge on health and safety since the Health and Safety at Work Act in 1974. As mentioned above there have been requirements in the Factories Act and other legislation to take personal precautions when handling harmful materials but this was too narrow and specific, e.g. the Chemical Works Regulations 1922.

The COSHH Regulations set out, for the first time, a strategy for safety with hazardous substances to cover all places of employment, processes and activities which could prove hazardous to health at work. As there are still many such places in existence today, nearly every employer, regardless of the size of their organisation, has to take some action to comply with the Regulations.

There are 19 Regulations in total but eight of these are mainly administrative in nature. The *primary* Regulation is no. 6, which requires employers to carry out an assessment of the risks presented by work which expose employees and others e.g. contractors, visitors, the public etc., to substances harmful to health. Again, these Regulations are not really new. Section 2(2)(b) of the Health and Safety at Work Act states that

> '... *an employer shall, so far as is reasonably practicable, ensure the safety and absence of risks to health in connection with the use, handling, storage and transport of articles and substances'.*

This is an example of extremely well drafted legislation. A moment's thought will confirm that there are no work activities which are outside the four categories of use, handling, storage or transport, and everything used at work is either an article or a substance. There are no exceptions. But like most of the Sections of the Act, the general nature of this requirement did not attract much attention until the COSHH Regulations appeared, expanding on this requirement in quite some detail, spelling out with some precision, exactly what s. 2(2)(b) required.

REGULATION 6 – RISK ASSESSMENT

This is the most meaningful of all 19 Regulations and states that

> '*an employer must not carry on work which is liable to expose any employee to any substance hazardous to health unless a*

> *suitable and sufficient assessment has been made of the risks created by that work to the health of those employees and of the steps which need to be taken to meet the requirements of the Regulations'.*

This should be clear enough. Risk assessment has been with us since 1975 when the Health and Safety at Work Act came into force. It was given a legal boost with the 1992 Management of Health and Safety at Work Regulations. Risk assessment and management has been publicised from every angle for years now and has received maximum coverage in professional and trade journals. It makes good business sense as risks are, in effect, threats to the assets of a business and ill health amongst employees is a costly, as well as a socially unacceptable, occurrence.

CARRYING OUT A COSHH RISK ASSESSMENT

As with traditional risk assessments a COSHH assessment follows a series of logical, progressive steps. These are not fixed; they can be designed by the assessor and will depend on the number of hazardous substances and the situations and processes in which they arise. The point is there must be a clear start and finish to the assessment which should begin with the identification of substances covered by the Regulations (the inventory creation) and ending with details of the procedures to monitor and review the assessments. A simple, straightforward method is outlined below.

Step one

Carry out an inspection of the entire premises noting down all substances which are highly toxic, toxic, harmful, irritant, carcinogenic, corrosive, oxidising and so on. Do not include flammable liquids, liquefied petroleum gases, lead or asbestos. These substances are covered by their own separate regulations. As you draw up this inventory note where the substances are used or produced, e.g. bought in, process produced, maintenance, raw materials and so on. This is necessary otherwise you may forget where you encountered an inventory item if the list is very long.

Step two

Collect as much information as possible on each item. This can be obtained from suppliers (they have to provide it). Place this information under relevant headings. These will cover such factors as chemical formula – important to differentiate between products known by their trade names; how much of the substance is used, produced, stored or despatched per annum – large or small quantities will pose different risks; how it is supplied, used or stored – in substantial packing or fragile containers; is information available from suppliers and it is adequate? Much suppliers' data contains irrelevant information such as 'closed cup boiling points' and other unnecessary detail.

Is an internal data sheet available containing the immediate, emergency action *only*, and is a copy of this posted where the material in question is likely to affect employees?

What form does the main hazard take? Is it a gas, vapour, solid, fume, liquid or dust? Everyone should be aware of this fact.

Finally, a summary of the hazard and its effect on the body should be known, e.g. 'Harmful: possible risks of irreversible effects through inhalation, in contact with skin and if swallowed'.

This is a vital step. It is imperative that as much information as possible about the substances comprising the inventory is known. Very often items with different trade names are the same substance but in different strengths. For instance, sodium hypochlorite, a strong, corrosive disinfectant and bleach containing chlorine, is sold under the trade name 'Hypo' in a concentrated form and 'Kleen Concentrate' in a greatly diluted form for cleaning canteen food surfaces and tables.

Step three

Having assembled the inventory of 'coshable' items and obtained the maximum amount of information possible you are now in a position to commence assessing the risks these items pose. For each item ascertain and write down what it is used for. Don't forget the substances used by the cleaners at night when everyone has gone home. Many cleaning materials are extremely dangerous; for example, 'toilet descaler' can be hydrochloric or phosphoric acid.

Make a note of the *present* controls in place. Sometime these are more than adequate and nothing else needs to be done. Far too many managers think that Regulations such as these require a whole new set of procedures and controls to be drawn up when existing measures are perfectly adequate.

Check whether the container has any pictogram (the orange square containing a black diagram and the legend 'toxic', 'irritant', etc. Note this fact down. Then consider the likely cause of exposure. Could it be an accidental spillage? Leakage? Overdosing? Giving off excessive vapour? And so on.

Who are the people most likely to be exposed, i.e. staff only, or staff and the public? In this respect it must be ascertained whether the substance has any known standard such as an Occupational Exposure Standard in parts per million or milligrams per cubic metre as designated in the HSE publication EH40. If the substance is capable of creating airborne contamination it must be ascertained whether air quality measurements are needed and how often. If they are required the results of such measurements must be retained.

It will be seen how making this series of simple checks will give the risk assessor a good idea of the risks posed by the substances on the inventory, and now that these *have* been assessed, the appropriate control measures can be considered.

Step four

Again looking at each item, action to be taken as a result of the risk assessment must be decided. This can probably range from 'nil' to 'need to install LEV system'. At this stage the range of control measures can be considered. These are, in hierarchical order:

- **Elimination** of the hazardous substance completely. Often difficult.
- **Substitution** of the substance by a less hazardous one. Possible but many substitutes are inferior to the original.
- **Enclosure** of the process or substance. Possible where the material can be kept in containment and does not have to be handled by people.
- **Isolation** poses the question – can it be put elsewhere? Can the space between the substance, process etc., be usefully increased? It should be remembered that this strategy can create maintenance problems.

- **Ventilation** is a feasible but often expensive control strategy. However, if the contaminant is essentially airborne, LEV is probably the most effective way of dealing with it.
- **Personal protective equipment,** when issued, signifies that everything that is reasonably practicable has been done to control exposure and that PPE is the last resort. When PPE forms a method of control see that it is specified by type.

Whatever control measures are decided upon, they will come under one or more of the above headings. The decisions regarding these measures must come from within the undertaking itself. This should not pose a problem but to some managers it does. It is difficult to comprehend the reasons why managers who are competent and experienced in controlling other corporate risks and hazards to their businesses cannot manage threats to employee health posed by hazardous substances used on their premises on a daily basis.

New or changed control measures will require revised operating procedures and in this context it is essential to ensure these are communicated to *everyone* concerned. Throughout history the root cause of many errors, disasters, incidents and accidents has proved to be a failure in communication. Make sure maintenance procedures are updated, too, where necessary.

Step five

This is the final stage in the COSHH assessment where monitoring and review procedures are established. The frequency of ongoing air quality testing, LEV test/ examination frequency and checks that correct systems of work are being followed must be set. Personal protective equipment, often left to look after itself, must be properly maintained and checked. New arrangements might require new training programmes; unlikely, perhaps but it is better to make sure. Finally the future of reassessments should be set, starting at frequent intervals and extending these as appropriate with experience.

Using this assessment programme is easy, straightforward and effective. Compared with some highly sophisticated and equally convoluted systems, it is clear and can be carried out by anyone after a short, familiarisation session. It will be seen that if necessary, certain steps can be allocated to different people, for example, the inventory creation and the knowledge gathering phases can be farmed out to separate members of staff. Even the procedures in each step are sufficiently discrete to be delegated to individuals.

One mistake which is often made is to carry out an overall assessment for an entire organisation in one fell swoop. This results in massive box files of data, where retrieval of information is difficult. Where an organisation comprises clearly identifiable discrete departments, each department should produce its own assessments and retain copies of these. Management may retain a master copy but the real value of these assessments lies in their being kept where the hazardous substances are located and used. These individual assessments can often be condensed into a single summary sheet as shown in table 8.1. However the detailed data sheets themselves should still form the principal assessment.

HEALTH SURVEILLANCE

This is an area where considerable misunderstanding exists. Where a company has substances on its premises which are subject to COSHH assessments it is *not* necessary for every employee exposed to these to have health surveillance. This procedure is only required under specific circumstances, for example

- where an employee is engaged in one of the processes listed in Schedule 5 of the Regulations, for example the manufacture of vinyl chloride monomer (VCM) *and is likely to receive significant exposure to the substance involved* (author's italics).
- where employees are exposed to a substance well known to be linked to a specific disease or adverse health effect, *and* there is a reasonable likelihood under the conditions of the work of that disease or effect occurring, *and* it is possible to detect the disease or adverse health effect (HSE's italics).

It will be seen, therefore, that the circumstances under which health surveillance is required are specific and quite reasonable, and they do not always require expensive medical assistance. Trained supervisors can check for dermatitis or ask questions about breathing problems where work involves substances know to cause asthma, and so on. Simple records should be kept (in hard copy) of any such health surveillance carried out and should comprise individual personal details and dates. Often this is all that is needed. Records do have to be retained for 40 years, and should an organisation cease trading the records must be deposited with the nearest HSE office.

Table 8.1 Sample COSHH Summary Sheet

S M Enterprise and Son Limited COSHH ASSESSMENT

Date ....

Substance	Chemical component	Use	Data Sheet?	COSHH hazardous	Exposure liklihood	Affected	Mode	Exposure limits/ Physical hazards	Air sampling	Safe system of work
Frekote 770-NC	Aliphatic hydrocarbon Dibutyl ether	Releasing agent	YES	YES Mildly toxic Low hazard rating	During releasing operations	Operatives/staff	Skin contact	1000ppm Fire hazard – emits fumes	N/A	Avoid skin contact. Wash off all liquid from skin
Coppergrease	Copper particle paste with synthetic grease	Anti-seize compound	YES	YES Skin and general irritant	During production and maintenance	Operatives/staff	Skin contact	None. Is an irritant. Oxidising agent	N/A	Avoid skin contact and observe good skin hygiene
Boss gas leak detector spray	Aqueous surfactants with CO_2 propellant	Gas leak detection	YES	NO	During checks for gas leaks	Maintenance staff	Skin contact/ inhalation	None – but avoid excessive skin contact	N/A	Avoid skin contact. Wear eye protection when using
Truklene	Blend of Na silicates; Na phosphates; Na salt of nitrilo acetic acid and detergents	Traffic film remover	YES	YES Skin and general irritant	During removal of traffic film from vehicle and other surfaces	Drivers and maintenance	Skin contact	None – but constant skin exposure will cause irritation/inflammation	N/A	Avoid excessive skin contact. Wear gloves and eye protection if needed
Descaling fluid	Phosphoric acid	Toilet cleaner	YES	YES Highly corrosive	Cleaning and descaling toilets	Cleaning staff	Skin and inhalation	None – but corrosive and dangerous if swallowed	N/A	Wear gloves and eye protection when using
7-132 cooling tower treatment	Microbicide cont. isothiazolones	Cooling tower treatment	YES	YES Corrosive and strong allergen	During cooling tower treatment	Contractor and maintenance	Skin, mouth inhalation	None – but causes irreversible eye damage, skin burns and dermatitis	N/A	Wear rubber gloves and apron, goggles and face shield
Graffiti remover K 14	N-Methyl – 2 – pyrollidone; pentane; n-butane; iso-butane; propane	Removal of ballpoint, crayon and aerosol paint	YES	YES Severe irritation on over-exposure	During the removal of graffiti and other similar vandalism	Operatives; passers by; maintenance	Skin, eye contact; inhalation	Minimum OES 100ppm spray is highly flammable	N/A	Use gloves and respiratory protection. Protect eyes
Yellow tabs	1, 4 Dichlorobenzene	Toilet cleaner	YES	YES Severe irritation from vapour and powder	Placing in toilet channels; cleaning toilet channels	Cleaning staff	Skin, eyes respiratory system	OES 25ppm for 1,4 Dichlorobenzene; irritant; high exposure can affect liver and kidneys	N/A	Use rubber gloves. Protect eyes
C Glass	Isopropanol; sodium dioctyl; sulpho succinate	Cleaning of glass and mirrors	YES	YES Mild irritant	During glass and mirror cleaning	Cleaning staff Maintenance	Skin, eyes	OES 400ppm for Isopropanol	N/A	Use lightweight rubber gloves. Protect eyes
Silicone spray	Hydrocarbon distillate; liquefied petroleum gas	Releasing agent	YES	YES Mild irritant	During releasing operations	Operatives	Skin, eyes respiratory system	OES 400ppm for hydrocarbon distillate	N/A	Use hand and eye protection. Ventilate work area
Pearl pink	Aqeous solution of surfactants and preservative	Washroom cleaning	YES	NO	N/A	N/A	N/A	NONE	N/A	None required
Clean and shine polish	Aliphatic hydrocarbon solvent blend; kerosene; butane; propane; silicone emulsion	Aerosol cleaner	YES	NO	N/A	N/A	N/A	NONE	N/A	Avoid skin contact; eye contact and inhalation
Amendments										

The COSHH Regulations are not the onerous burden some employers feel they are, but they do represent good practice. It is reasonable enough to keep employees safe from over-exposure to any substances hazardous to their health on their employer's premises. Absences due to ill health are expensive and can result in costly compensation claims. The law requires employers to know about such substances; assessing the risks they pose, via a COSHH assessment makes good sense all round.

THE CHEMICALS (HAZARD INFORMATION AND PACKAGING FOR SUPPLY) REGULATIONS 1994

These Regulations, known colloquially as the 'CHIP' Regulations, came into force in 1995. Their purpose is to classify substances and preparations which are hazardous for supply based on their physicochemical properties, effects on people's health and on the environment. The classification enables identification of hazardous properties relevant to normal handling and use. The Regulations require substances so classified to display a supply label which is designed to convey essential information about the product so that the user can draw up the necessary precautionary measures for its use. This label does not replace the suppliers' safety data sheets; rather it complements them and can be instrumental in drawing the users' attention to more comprehensive information on those sheets.

The label must take into account all potential hazards likely to arise in normal handling and use – although the latter requirement is rather wide ranging. Information must also be provided on the label regarding the name, full address and telephone number of a person who is responsible for supplying the substance, thus making available the correct information on what action is to be taken in an emergency.

The CHIP Regulations introduced standard 'phrases' giving information on risk, safety, combinations of particular risks, and combinations of safety precautions. These phrases are particularly useful as they are specific, concise and enable COSHH assessments in particular to be drawn up with a degree of standardisation. Examples of these phrases are as follows:

Risk phrases: indication of particular risks

R14   Reacts violently with water
R24   Toxic in contact with skin
R41   Risk of serious damage to eyes.

Combination of particular risks

R26/27/28 Very toxic by inhalation, in contact with skin and if swallowed
R26/28   Very toxic by inhalation and if swallowed
R36/38   Irritating to eyes and skin.

Safety phrases: indication of safety precautions

S4   Keep away from living quarters
S17   Keep away from combustible material
S46   If swallowed seek medical advice immediately and show this container or label.

Combination safety phrases

S3/9/14 Keep in a cool, well ventilated place away from ...(incompatible materials to be indicated by manufacturer)

S36/39   Wear suitable protective clothing and eye/face protection

S20/21   When using do not eat, drink or smoke.

It must be pointed out that the CHIP Regulations place duties on *manufacturers* and *suppliers* of hazardous substances for use at work rather than the *users*, e.g. the SME. They establish the concept of a supply *chain* and emphasise the dependency of all those in the chain, including the end user, on accurate health and safety information. For the end user such information is essential for carrying out practical risk assessments. In short, the CHIP Regulations 'lean' on the suppliers to provide more information about their products than they might feel inclined to provide on their home-produced health and safety data sheets.

The main duties, then, on manufacturers and suppliers, are as follows:

- to **classify** substances in accordance with Approved Supply List. This List indicates the specific classification (under the classifications of physico-chemical, health or environment) and labelling requirements for over 1,400 substances that are agreed with the EU.
- to **clearly label** substances in accordance with standard classifications to which Risk and Safety Phrases are applicable.
- to use the standard Risk Phrases and Safety Phrases described above in addition to standard warning signs.
- to provide health and safety **data sheets** in addition to the above information.
- to ensure hazardous substances are *packaged* in a safe manner. Packaging means, in plain English, keeping the hazardous contents of the package in containment, preventing leaks or seepage and rendering the package safe during all handling operations until it is finally opened. The contents of the package shall not escape.

As these Regulations are intended primarily for manufacturers and suppliers only, an outline of the main points is given in this chapter. However, for information and interest, the categories of danger in Schedule 1 of the CHIP Regulations are reproduced below. They are quite comprehensive and the SME manager need look no further when trying to categorise hazardous substances during the carrying out of a COSHH assessment.

PHYSICOCHEMICAL PROPERTIES

- Explosive; oxidising (gives rise to exothermic reactions); extremely flammable
- Highly flammable and flammable.

HEALTH EFFECTS

- Very toxic (very small quantities can be fatal); toxic (small quantities can be fatal)
- Harmful (can cause death when inhaled, swallowed or absorbed through the skin)
- Corrosive; irritant (can lead to inflammation, especially in the mucous membranes of the body e.g. eyes, nose, mouth); sensitising; carcinogenic; mutagenic; toxic for reproduction (teratogenic, e.g. the drug thalidomide).

DANGEROUS FOR THE ENVIRONMENT

Substances which, if they were to enter the environment, would probably cause harm to one or more components of the environment.

SUMMARY

This chapter on legal aspects of occupational health is intended primarily to provide basic information for the SME with particular emphasis on compliance with the COSHH Regulations from the point of view of the workplace and those in it. The chapter is not intended as a definitive work on the law and for that reason dwells in the main on the detail the SME manager needs to know.

The section on carrying out a COSHH assessment is dealt with in some detail and offers a simple, straightforward and relevant model on which a manager can carry out assessments in the minimum time, at the same time deriving maximum benefit. Some systems on offer are unnecessarily complex and unduly detailed and much time is wasted trying to understand them.

The abridged section on the CHIP Regulations is primarily for information to let the SME manager know that there is a clear, legal requirement for suppliers of substances hazardous to health to state in clear detail precisely what the dangers of their products are. The days have long gone when suppliers could get away with vague generalisations about their products such as 'Can cause irritation if inhaled. Use in a well ventilated area' without specifying what the causative agent of the irritation was or what constituted a 'well ventilated area'.

CHAPTER 9

Personal protective equipment

Personal protective equipment, mainly respiratory protection, has been used in some form or other since ancient times.

INTRODUCTION

The Romans devised crude face masks made out of pigs' bladders for their slaves to wear when mining various minerals and metals supposedly to protect them from the dust of mining operations. In more recent times workers in dusty processes have used equally useless 'protection' in the form of wet handkerchiefs tied across the mouth and, hopefully, the nose, to protect from the dust.

THE PERSONAL PROTECTIVE EQUIPMENT AT WORK REGULATIONS 1992

These Regulations were intended to correct a situation which had been unsatisfactory for many years notwithstanding legislation under the Factories Act and other statutes, orders and regulations. They set down specific requirements relating to provision, compatibility, assessment (of the suitability of PPE for the job in question), maintenance (of PPE in a serviceable condition) and the provision of suitable accommodation for the equipment. No longer was it acceptable just to provide a dirty, opaque pair of goggles hanging from a rusty nail in the wall above the abrasive wheels or a paint-covered respirator on a similar nail by the paint spraying booth.

'Personal' in the title of the Regulations has two meanings. Protective equipment is personal in that it is for the protection of the person (rather than a machine) and it is personal on an individual basis, i.e., respiratory protective equipment, hard hats, ear defenders and so on should not be shared. Each employee requiring such items should be issued with their own equipment for health and hygiene reasons, and must be responsible for maintaining such equipment in a serviceable condition.

GETTING EMPLOYEES TO USE PPE

It is a well known fact that the more one encumbers the human body with clothing, equipment, protection and so forth the more uncomfortable the wearer becomes. Also, the sensitivity to and awareness of proximate objects is very considerably reduced. This phenomenon is particularly noticeable in the case of the hands; employees are even reluctant to use barrier creams to protect the skin and, it would appear, are perfectly happy to risk contracting dermatitis – and then sue their employer because the condition can prevent them working. Time and time again one hears the well-worn refrain from delegates at health and safety gatherings, 'We provide all the PPE necessary but 'they' won't use/wear it'. The Regulations do not advise the employer what to do in such circumstances.

However, the law is quite clear on this subject. If, after a risk assessment, the employer is of the genuine opinion that an employee must use an item of PPE for that employee's personal safety and health while at work then the employee must comply. Of course, employers must show that they have done all that is reasonably practicable to minimise the risk before deciding on PPE as a last resort. In principle, the employer has a *bona fide* case for dismissing an employee who refuses to use PPE.

PROBLEMS CREATED BY PPE

The Regulations make a number of extremely pertinent points *against* the use of PPE which are worth reproducing here. Firstly, PPE protects only the person using it whereas controlling the risk at source protects everyone at the workplace. Secondly, theoretical maximum levels of protection are seldom achieved with PPE in practice and the *actual* levels are hard to assess. Thirdly, effective protection is only achieved with fully suitable PPE, correctly fitted and maintained and properly used. The point about restricting mobility and reducing sensitivity and awareness has already been made.

Training in the use of all types of respiratory PPE is essential even if the employees think they know it all. Facial hair, spectacles and some face shapes cause problems of leakage with full face masks or even ori-nasal respirators. The latter type of respirator is usually an effective fit and functions well but is often made of rubber type compounds which produce a disagreeable smell. The tendency is to settle for the white gauze type ori-nasal masks but there are often problems of leakage with these devices which do not necessarily arise in the rubber type products.

All these problems create difficulties, cause time to be lost and can become very expensive. Also, remember that no charge whatsoever may be levied on employees in respect of PPE under any circumstances. Again, these factors point to the preference to control the risk at source.

WHAT PPE IS REQUIRED?

PPE for use with hazardous substances includes the following:

- respiratory protective equipment (there is a hierarchy for this; see below)
- hand protection
- skin protection
- eye and face protection.

If one glances through the catalogue of one of the many reputable suppliers of PPE, the first thing that impresses – or confuses – is the vast range of equipment available. This is especially true in the case of respirators and the range of cartridge filters, and gloves. Although suppliers usually have very knowledgeable staff it is prudent for the employer to become acquainted with the type of material the hand protection required should comprise; the type of cartridge filter the ori-nasal respiratory masks require and the nature of the barrier cream that should be supplied. Some years ago, the author came across a cupboard full of dust masks sold to an engineering concern by a commission-happy salesman which were totally unsuitable for protecting against the respirable dust used in the company's processes.

There are two distinct methods of providing personal protection against contaminated atmospheres:

1 By purifying the air breathed

The inhaled air is drawn through a medium which removes the harmful substances, the nature of the medium (filter) depending on the contaminant – which must be known. These devices are known as *respirators.* Note that respirators do not provide air and therefore must not be used in oxygen deficient atmospheres.

Some different types of respirators are:

Respirators for dusts
- General purpose dust respirators
- Positive pressure powered (battery worn by wearer) respirators – hood or blouse
- High efficiency dust respirator.

For gases
- Canister type full face respirators
- Cartridge type full or half face respirators
- Emergency escape type e.g. the 'self rescuer' worn by miners.

For gases with dusts
- Canister type with particulate filter
- Cartridge type respirators.

2 By supplying air or oxygen from an uncontaminated source

The inhaled air is conveyed by air line or alternatively from cylinders carried by the person at risk. A device such as this is known as breathing apparatus (BA). Some types of BA are:

Self contained BA
- Closed circuit – without access to atmosphere. Wearer breathes same air over and over again
- Open circuit – face piece connected to cylinder with finite supply of air
- Emergency escape – specialised design.

Air line BA
- Air supplied through fresh air hose
- Air supplied through compressor air line.

The hierarchy of respiratory protective equipment mentioned above is, therefore, as follows:

- enclosed breathing apparatus (BA)
- service fed BA, i.e. via a fresh air hose or compressor air line
- battery powered filtering respirator
- cartridge respirator
- filtering face piece.

Note: in a positive pressure face piece the pressure inside the face piece is slightly higher than the outside atmospheric pressure so any leakages are outwards from the face piece.

RESPIRATORY PROTECTIVE EQUIPMENT FOR SMES

In the majority of cases and for most SMEs respiratory protective equipment (RPE) will be the norm and will be in the form of cartridge or canister type respirators of the ori-nasal type. Dust respirators account for by far the largest number of devices in use industrially to protect workers against harmful dusts, many of which are not only irritant, but, as explained previously, can lead to illness or disease. In addition there are available lightweight, simple face masks of many types and makes to protect workers against *nuisance* i.e. non-respirable dusts, and non-toxic sprays. These do not offer any real measure of protection, though, and are more a placebo than anything else. Cartridge type ori-nasal respirators with the correct filter give protection against *low concentrations* of mildly toxic gases, but for real protection against larger concentrations of more toxic gases, canister respirators must be used, failing which full BA is necessary.

SUMMARY

This final chapter has provided a brief overview of the types of personal protective equipment which can be considered for protecting workers in contaminated work environments. As inhalation of contaminants is the greatest single risk facing workers in such situations, most of the emphasis has been on respiratory protective equipment. If this equipment is needed – as a last resort – then it should be selected with care and the employees concerned should be consulted on choice and type. Time must be set aside for training in its use and maintenance. Make full use of the PPE supplier's expertise and if they do not inspire confidence, approach a competitor. Most suppliers provide basic training and instruction in the use of their products – free of charge and on your premises. Make full use of such facilities.

Index